INTERNATIONAL CENTRE FOR MECHANICAL SCIENCES

COURSES AND LECTURES - No. 203

DONALD L. DEAN
NORTH CAROLINA STATE UNIVERSITY
RALEIGH, NORTH CAROLINA

DISCRETE FIELD ANALYSIS OF STRUCTURAL SYSTEMS

SPRINGER-VERLAG WIEN GMBH

Originally published by Springer-Verlag Wien New York in 1976

ISBN 978-3-211-81377-5 ISBN 978-3-7091-4360-5 (eBook)
DOI 10.1007/978-3-7091-4360-5

LECTURE I

INTRODUCTION TO DISCRETE FIELD MECHANICS

Course Goals. – These lecture notes were prepared for publication prior to the lecture series and with limited knowledge as to the background of the typical participant. It was assumed, therefore, that the audience would have backgrounds in mathematics through differential equations, including introductory work in partial differential equations, and backgrounds in structural theory that include advanced frame analysis and introductory work in plate and shell theory. The course goals will be somewhat flexible depending upon the lecturer's observation of the rate at which the new material can be assimilated by the audience, but planning for the lecture series was based upon the ambitious goal of both introducing the fundamental concepts of discrete field mechanics and covering work at the frontiers of present knowledge of the subject as related to structural analysis.

The term discrete field analysis denotes the body of concepts used to obtain field or functional solutions for systems most accurately represented as a lattice or a pattern of elements. The mathematical model for a lattice is characterized by the use of discrete (as opposed to continuous) variables for at least one of the coordinates required to cover the field. The application of discrete field methods to the analysis of structural systems is not new. Although there have been more recent publications(*), the best existing book on the subject(**) was written more than 45 years ago. It is somewhat surprising, therefore, to find that the state of the art is relatively undeveloped, especially as compared with continuum mechanics. To emphasize this point, one might observe that the student completing this short course should be able to derive formulas that could be duplicated by only a few dozen other structural analysts in the world today. It is the author's hope that the acquisition of this unique capability by the number of additional workers in the audience will significantly increase the rate at which the state of the art of discrete field mechanics is being advanced. It is also his belief that a more wide-spread capability in discrete field methods will serve to advance structural design work by encouraging increased use of exotic lattice and composite lattice-continuum systems. The derivation of discrete field formulas for the design of complex structures often requires more of a theoretical background than that possessed by the average practitioner; however, once the formulas have been derived they can be used by most degree trained designers and a catalogue of the available solutions would be quite a useful addition to the designers' literature. It will not be the object here, however, to start such a catalogue. Where a choice must be made between an illustrative problem of great design utility and a simpler one that more clearly illustrates the principles and techniques, the latter case will be selected. It is hoped that the sample problems will have heuristic value in leading the participants to apply the methods to practical design examples which can be collected and catalogued later.

(*) Structural Analysis by Finite Difference Calculus, by T. Wah and L.R. Calcote, Van Nostrand Reinhold, New York, 1970.

(**) Die Gewohnlichem and Partiellen Differenzengleichungen der Baustatik, by Fr. Bleich and E. Melan, Julius Springer, Berlin, 1927.

Comparison with Other Methods. – Discrete field approaches are logical successors to matrix and simultaneous equation approaches to the rational analysis of large structural lattices having a regular pattern of identical elements. The typical saving in computational effort (or expense in the case of automatic computers) in using a discrete field solution over a matrix approach for the analysis of a structural system of many elements (e.g., eight or more) is from one-half to four orders of magnitude; i.e., the matrix method requires from 3 to 10,000 times as many computations. The comparison is all the more impressive if one also takes into account the massive set up effort and possibility of significant and accumulative round-off errors attendant to the matrix analysis of a large system.

The substitute or "equivalent" continuum method is another popular alternative to the matrix method for analyzing structural lattices. It shares with the discrete field methods the advantages inherent to field techniques in that the solution does not become more cumbersome as the size or number of elements in the system increases. The disadvantage is the lack of rationality, at least in selecting an "equivalent" continuum and in interpreting the continuous field results for application to the discrete field. Furthermore, the empirically derived continuous field model is often no more tractable than would be the rationally based discrete field model.

For mixed discret-continuous systems, such as a composite grid-plate or "waffle" plate system, the discrete field method is the only rational approach available. That is, the matrix approach is not a viable alternative as general boundary solutions are unavailable for the element of a multidimensional lattice with multidimensional elements. The finite element method, for which an error analysis is not yet available, is often used as a semi-empirical approach to the analysis of such structures. The finite element method may be viewed as a procedure whereby the real system for example a mixed discrete-continuous structure, is replaced by an approximating discrete system which is then solved by matrix methods. It may be possible to rationalize the error analysis of the finite element method through comparison of exact field solutions to the real and the substitute discrete models. Another related open problem is the investigation of the use of field solutions to regular lattices as approximations to the solution for slightly irregular lattices.

Difference Equation Models. – The only discrete field method in wide use at present is one in which the system model is written as a difference equation plus boundary conditions. Such an approach is herein termed a micro approach. The method is as follows:

1. general boundary solutions are found for the typical basic element between joints which are labeled by discrete coordinates:
2. the mathematical model, in the form of a difference equation, is derived by relating the element behavior to that of adjoining or connecting elements so as to satisfy equilibrium and compatibility at the joints (The difference equation may be written either as a recurrence relation or by use of finite difference operators.);
3. closed form functional solutions (which satisfy the governing difference equation and the boundary conditions) are then found for the joint quantities; and
4. the joint quantities can then be substituted back into step 1. to find the desired mid-span quantities. For example, the micro-flexibility approach to the analysis of a prismatic beam continuous over evenly spaced unyielding suppports leads to the following well-known "three moment" recurrence relation – i.e., difference equation:

$$M(r+1) + 4M(r) + M(r-1) = -6\frac{B}{\ell}\Phi(r) \tag{1}$$

in which M(r) is the joint moment at the node denoted by the discrete coordinate r, B equals flexural rigidity, ℓ equals node spacing and Φ (r) is the slope discontinuity occuring at joint r with moment releases or hinges introduced at joints r–1, r, and r +1. Boundary conditions are required to complete the model . For example, if the beam is simply supported at r = 0 and fixed at r = n the boundary conditions are :

$$M(0) = 0 \qquad 2M(n) + M(n-1) = -\frac{6B}{\ell}\Phi(n) \tag{2a,b}$$

It should be noted that the use of Eqs. 1 and 2 to generate simultaneous equations yields a n–1 x n–1 banded or "tri-diagonal" coefficient matrix.

The difference model for the same beam continuous over identical spring supports is as follows:

$$\begin{bmatrix} \frac{\ell}{6B}(\Delta_r + 6) & \frac{1}{\ell}\Delta_r \\ \frac{1}{\ell}\Delta_r & -\sigma \end{bmatrix} \begin{bmatrix} M(r) \\ w(r) \end{bmatrix} = \begin{bmatrix} -\Phi(r) \\ -P^e(r) \end{bmatrix} \quad (3a)$$

$$\begin{bmatrix} 1 & 0 \\ 0 & 1 \end{bmatrix} \begin{bmatrix} M(0) \\ w(0) \end{bmatrix} = \begin{bmatrix} 0 \\ 0 \end{bmatrix} \quad (3b)$$

$$\begin{bmatrix} \frac{\ell}{6B}(3 - \nabla_r) & 0 \\ 0 & 1 \end{bmatrix} \begin{bmatrix} M(n) \\ w(n) \end{bmatrix} = \begin{bmatrix} -\Phi(n) \\ 0 \end{bmatrix} \quad (3c)$$

in which Δ_r denotes the second central difference operator, ∇_r denotes the first backward difference operator, σ equals the support spring constant and $P^e(r)$ equals the statically equivalent joint load at r due to loads in the two contiguous spans. Note that this matrix difference equation for the vector joint quantities $\{M(r), w(r)\}$ could be used to generate simultaneous equations having a doubly banded coefficient matrix.

Summation Equation Models. – A powerful alternative to the micro or difference equation method of discrete field analysis, the macro or summation equation method, has recently become available(*). The term macro refers to the fact that the method is based upon general load solutions to major system components, whose span dimensions correspond to those of the whole system, as opposed to the solutions for the between-nodes system elements which form the basis of the micro approach. The macro method can be outlined as follows:

(*)"Macro Approach to Discrete Field Analysis", by D.L. Dean and H.V.S. GangaRao, Journal of the Engineering Mechanics Division, ASCE Proceedings, August 1970.

1. general load solutions (usually Kernel functions) are found for the system components which satisfy the given system boundary conditions;
2. the mathematical model, in the form of a summation equation, is written by expressing compatibility between system components in terms of unknown interactive forces;
3. closed form solutions which satisfy the governing summation equation, are then found – usually in the form of finite sinusoidal series – for the unknown interactive force field; and
4. this solution for the node forces can be substituted into the expressions found in step 2 to find other node quantities or in step 1 to find the entire field of component descriptors – i.e., continuous functions valid at nodes and between nodes.

Attention is called to the fact that the macro approach does not require a general boundary solution for the basic element; thus, some multidimensional lattices with multidimensional elements, such as a "waffle" plate with tractable external boundary conditions, can be analyzed by the macro approach but not by the micro approach. On the other hand, a system with tractable element boundary conditions, but intractable components, such as a continuous beam with repetitively tapered between node spans, can be analyzed by the micro approach but not by the macro approach.

The macro-flexibility approach to the analysis of a prismatic beam continuous over evenly spaced spring supports leads to the following model:

$$w(x) = w^0(x) - \sum_{\alpha=1}^{n-1} R(\alpha) K\left(x, \frac{L}{n}\alpha\right) \tag{4a}$$

$$\frac{1}{\sigma} R(r) = w^0\left(\frac{L}{n} r\right) - \sum_{\alpha=1}^{n-1} R(\alpha) K\left(\frac{L}{n} r, \frac{L}{n}\alpha\right) \tag{4b}$$

in which σ equals the support spring constant, $R(r)$ denotes the field of unknown intermediate reaction forces, $r = 1,(1),n-1$, $w^0(x)$ equals the beam deflection for $R(r) = 0$, $K(x,\xi)$ equals the beam deflection Kernel function at x due to a unit impulse load at $x = \xi$ for the given external boundary conditions, and L equals the total beam length, i.e., $n\ell = L$. Note that here the model is a single summation equation, Eq. 4b, instead of the two difference equations plus boundary conditions, Eqs. 3, required for a micro approach. (The summation model for unyielding supports is covered by setting $1/\sigma = 0$). Also, use of the macro model to generate simultaneous equations leads to a full n–1 x n–1 coefficient matrix instead of the doubly banded (2n – 2) x (2n – 2) matrix generated by the micro model. Another significant distinction is that the macro solution yields results for the entire continuous field $0 \leqslant x \leqslant L$ whereas the micro approach yields only node values for r = 1, (1), n–1 and required a separate step to find any desired between node quantities.

Relation to Numerical Analysis. – Due to the current popularity of various finite difference and finite element methods (some of which employ elements of the calculus of finite difference) in structural mechanics, a brief comparison between these numerical methods and discrete field methods is indicated. A succinct statement of the distinction between the principal subject of this course and these numerical methods is as follows: Finite difference and finite element methods are used to find approximate open form numerical solutions for continuum models whereas discrete field methods are used to find exact closed form solutions for discrete models. The numerical analyst starts with a continuum approximation to the real physical system which for some reason, usually irregularly varying coefficients, is intractable by closed form methods. He then replaces the continuous model by an approximating discrete model. The finite difference approach requires that the various differential operators be replaced by a truncated series of finite difference operators while the finite element approach requires that one arrive at the substitute discrete model through use of an approximate boundary solution for the structural elements to write a difference or recurrence equation as in the micro field approach. In either case, the continuous model is replaced by an approximating discrete model which can be shown to comprise an exact model for a physically identifiable discrete system. Some workers do this directly on strictly physical grounds and refer to the approach as a substitute frame or substitute truss method. In all instances, however, the numerical analyst is seeking a discrete model for the purpose of getting approximate node answers through some simultaneous equation solution technique as opposed to seeking a closed analysis of the approximating discrete model. It would seem that use of a closed form lattice analysis by an experienced discrete field worker would be a promising approach for reconciling the discrepancies between the finite difference, finite element, and substitute frame methods for numerical analysis and for establishing the rational error analysis which are so far lacking in the latter two.

LECTURE II

SOLUTIONS TO DIFFERENCE EQUATIONS

Walk-Thru Numerical Solution to Ordinary Difference Equations. – The primary goal of the course is the closed analysis of regular discrete models. An irregular discrete model can be used to generate simultaneous equations for one of the standard numerical procedures and this is usually done; however, the field concept can be very useful here especially in the case of models with several dependent variables. For those difference models that are too irregular to be tractable by closed form techniques, a numerical walk-thru procedure is often a much more efficient alternative to the simultaneous equations or matrix approach. As an example of an irregular difference model, consider the general three moment equation; i.e.,

$$\left(\frac{\beta \ell}{B}\right)_r M(r+1) + \left[\left(\frac{\alpha \ell}{B}\right)_r + \left(\frac{\alpha' \ell}{B}\right)_{r-1}\right] M(r) + \left(\frac{\beta \ell}{B}\right)_{r-1} M(r-1) = -\Phi(r)$$

(5a)

in which the typical span boundary flexibility coefficients are given by

$$\text{(5b)} \qquad \begin{bmatrix} \phi \\ \phi' \end{bmatrix} = \left(\frac{\ell}{B}\right)_r \begin{bmatrix} \alpha & -\beta \\ -\beta & \alpha' \end{bmatrix}_r \begin{bmatrix} M \\ M' \end{bmatrix}_r + \begin{bmatrix} \theta^s \\ \theta'^s \end{bmatrix}_r$$

where ϕ and ϕ' are the left and right end rotations with respect to the deformed chord (equal to the end rotation for the case of unyielding supports); θ^s and θ'^s are the corresponding end rotations due to midspan loads with simple end supports ; $\Phi(r) = \theta^s_r - \theta'^s_{r-1}$; and $M(r) = -M'(r-1)$. Equation 5a is in the most convenient form for a walk-thru solution, but it is of passing interest to show the equation in operator form for the case where all spans are prismatic; i.e., $\alpha_r = \alpha'_r = 1/3$, $\beta_r = 1/6$

$$\text{(6)} \qquad \nabla_r \left[\left(\frac{\ell}{B}\right)_r \Delta_r M(r)\right] + 6\left[\mu_r \left(\frac{\ell}{B}\right)_r\right] M(r) = -6\,\Phi(r)$$

in which ∇_r, Δ_r and μ_r are the backward difference, forward difference and backward mean operators respectively (see p. 93).

The most efficent algorithm for obtaining an open form solution to an irregular second order difference equation such as Eq. 5a is to numerically walk thru a particular and a homogeneous solution. (In some cases, use of banded matrix techniques equals walk-thru efficiency from the standpoint of number of computations but are less convenient to set up). It is possible to program such a solution for general boundary conditions; however, it is usually preferable to design the algorithm for specific boundary conditions, e.g., $M(0) = M(n) = 0$, and add pseudo end spans with the appropriate properties to simulate other conditions if required; e.g., for a fixed condition at $r = 8$, add a pseudo end span which is hinged at $r = 9$ and has a L/B that is several orders of magnitude smaller than the stiffest real span. Thus, unusual boundary conditions are dealt with as additional input data for the element bean spans.

An outline of the walk-thru procedure, as applied to the continuous beam problem, is as follows:
1) walk thru a particular solution which satisfies the conditions $M^P(0) = M(0)$ and $M^P(1) = 0$,;
2) walk thru a homogeneous solution, $\Phi(r) = 0$, which satisfies the conditions $M^h(0) = 0$ and $M^h(1) = 1$; and
3) compute the constant in the total solution, $M(r) = M^P(r) + CM^h(r)$, by use of the given value for $M(n)$ — with these initial conditions, $C = M(1)$. A numerical example of the solution for a model consisting of Eq. 6 plus the boundary conditions $M(0) = -10k-ft.$ and $\theta(8) = 0$ is shown in Table 1.

TABLE 1
Walk Thru Solution to Irregular Continuous Beam Cantilevered at
$r = 0$ and Fixed at $r = 8$

Joint r	$(L/B)_r$ kip.-ft.	$\Phi(r)$	$M^P(r)$ kip.-ft.	$M^h(r)$	$M(r)$ kip.-ft.
0		--	-10.0	0	-10.0
	1.0×10^{-3}				
1		60×10^{-3}	0	1	-113.486
	$.3 \times 10^{-3}$				
2		50×10^{-3}	$-1,666.6\cdots$	$-8.666\cdots$	-183.121

	$.3\times10^{-3}$				
3		55×10^{-3}	3,666.6···	33.666···	-154.031
	$.4\times10^{-3}$				
4		60×10^{-3}	-12,783.3··	-111.33··	-148.550
	$.4\times10^{-3}$				
5		65×10^{-3}	46,566.6··	411.66··	-151.767
	$.5\times10^{-3}$				
6		60×10^{-3}	-158,193.3	-1,392.93	-114.798
	$.5\times10^{-3}$				
7		70×10^{-3}	585,486.67	5,160.06	-109.040
	1.0×10^{-3}				
8		30×10^{-3}	-1,677,783.3	-14,783.73	-35.480
	$[1.0\times10^{-9}]$				
9		--	$2,769,900\times10^{6}$	$24,407.4\times10^{6}$	0

It should be noted that the procedure may be subject to significant round-off errors. This problem can be detected through errors in satisfaction of the terminal boundary condition and corrected by recycling the solution so as to minimize the contribution of the homogeneous solution; that is, by changing the particular solution to $M^p(0) = M(0)$ and $M^p(1) = M(1)$ as $M(1)$ is known to a high degree of accuracy from the first cycle. Thus, the new particular solution is very nearly a total solution and the homogeneous solution is required only to correct for any remaining round-off errors. The selections of initial conditions such that $M^p(r) + CM^h(r)$ satisfies the initial boundary conditions independent of C permits the total solution to be written with only one homogeneous solution.

The above procedure can be easily generalized to cover the cases of higher order scalar difference equations (i.e., a single dependent variable) as well as vector difference equations (i.e., models with several dependent variables or simultaneous difference equations). An example of the latter is indicated in Eq. 7 below.

$$A_r \bar{X}_{r+1} + B_r \bar{X}_r + A_{r-1} \bar{X}_{r-1} = \bar{P}_r \tag{7}$$

in which A_r, B_r and A_{r-1} are n x n coefficient matrices and $\bar{X}_r$ is a n x 1 column matrix; i.e., a vector of unknown functions of r. Assuming that there are m initial and m terminal boundary conditions, the solution algorithm is as follows:

1) Step through a particular solution $\bar{X}_r^p$ which satisfies the given initial conditions, Eq. 7, and the m arbitrary conditions necessary to solve for $\bar{X}_2^p$ with r = 1 (each step requires premultiplication by A_r^{-1});

2) Step through m independent homogeneous solutions $\bar{X}_r^1, \bar{X}_r^2 \ldots \bar{X}_r^m$ (for $\bar{P}_r = 0$) each of which satisfies homogeneous initial conditions analogous to the given initial conditions and m additional arbitrary conditions selected so as to make the solutions

independent (e.g., $\bar{X}_1^1 = 1,0,0,\ldots.0, \bar{X}_1^2 = 0,1,0,\ldots0, \bar{X}_1^3 = 0,0,1,\ldots0$, etc.) and

3) determine the C's in the total solution $\bar{X}_r^t = \bar{X}_r^p + C_1\bar{X}_r^1 + C_2\bar{X}_r^2 + \ldots C_m\bar{X}_r^m$ so that it satisfies the terminal boundary conditions. The procedure is closely analogous to the scalar example shown in Table 1 and may be readily programmed for automatic computation. Here too it is preferable to program for given boundary conditions and then to accommodate different conditions by adding pseudo end spans which have the appropriate properties to simulate the actual boundary conditions.

Walk-Thru Numerical Solution to Partial Difference Equations. – It is also possible and practical to walk thru numerical solutions to difference equations with two or more independent variables or coordinates. The algorithm is similar to that for ordinary difference models. The present state of the art in walking thru partial difference models is such that one needs a number of homogeneous solutions equal to the conditions given for each node on the terminal boundary curve (or line) multiplied by the number of nodes on that boundary. This is a considerable saving over solving a number of equations equal to the number of unknowns multiplied by the total number of nodes; however, it may be possible to improve the technique in the future by working with unknown terminal boundary functions instead of boundary unknowns and thus to reduce the number of required homogeneous solutions by a factor equal to the number of terminal boundary nodes.

As an object problem for discussion of the walk-thru method for solving partial difference models, consider the equation for the equilibrium displacements of a doubly threaded net (*)

$$R_s \nabla_r \left[\frac{1}{a_r} \Delta_r w(r,s)\right] + S_r \nabla_s \left[\frac{1}{b_s} \Delta_s w(r,s)\right] = -P(r,s) \qquad (8a)$$

or

$$\frac{R_s}{a_r} w(r+1,s) + \frac{R_s}{a_{r-1}} w(r-1,s) + \frac{S_r}{b_s} w(r,s+1) + \frac{S_r}{b_{s-1}} w(r,s-1) -$$

$$-\left[\frac{R_s}{a_r} + \frac{R_s}{a_{r-1}} + \frac{S_r}{b_s} + \frac{S_r}{b_{s-1}}\right] w(r,s) = -P(r,s) \qquad (8b)$$

(*) Analysis of Structural Nets", Donald L. Dean and Celina P. Ugarte, Publications International Association for Bridge and Structural Engineering, Vol. 23, December 1963, pp. 71-90.

in which R_s are the tensions in the cables parallel to the r coordinates, S_r are the corresponding values for the other set of cables, a_r are the node spacings along the r axis and b_s are the node spacings along the s axis (see Fig. 1 for sketch with a_r and b_r constant). The following explanation of the walk-thru procedure refers to Eq. 8b and Fig. 1 but it should be understood that the method is applicable to any linear partial difference model. Step 1 – walk thru a particular solution, $w^p(r,s)$, which satisfies (a) the real initial conditions on the nodes marked by the dark circles (here the method will be to walk thru in the r direction but it could as well have been in the s direction), and (b) the assumed values for the first interior nodes marked by open circles; i.e., solve for w(2,s) using known w(0,s) and assumed w(1,s), increase r by one to find w(3,s), etc. until the entire field $w^p(r,s)$ is determined, including values for nodes along the terminal boundary (which will usually not satisfy the given terminal boundary condition). Step 2 – walk thru a number of independent homogeneous solutions (P(r,s) = 0) equal to the number of assumed values required in the particular solution by making the appropriate changes in the values assumed for the nodes marked by the open circles. (The homogeneous solution, as before, must satisfy the homogeneous initial conditions analogous (zero load terms) to the real initial conditions. Here the required number of homogeneous solutions is k = n – 1. If m were a smaller number, the walk-thru's would have been in the s direction so that only m – 1 homogeneous solutions would have been required. Step 3 – find the C's in the total solution so that it satisfies the terminal boundary conditions.

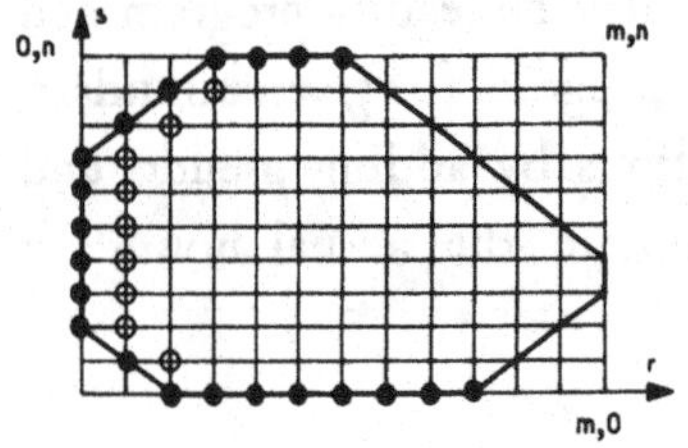

Fig. 1. Doubly Threaded Net with Arbitrary Boundary

$$(9) \qquad w(r,s) = w^p(r,s) + C_1 w_1^h(r,s) + C_2 w_2^h(r,s) + \dots C_k w_k^h(r,s)$$

Closed Form Analytical Solutions to Ordinary Difference Equations. – The object of this section and Appendices I and II is to briefly review some of the methods for writing closed form analytical solutions to ordinary difference equations with constant coefficients and to tabulate the results of certain operations that are frequently encountered in practice.

As a first example of writing a closed form solution to a difference equation, consider the case of a continuous beam with the same symmetric loading in each span and zero end moments; i.e., the model (see Eq. 1) is

$$(\Delta_r + 6)M(r) = -M_o \tag{10a}$$

$$M(0) = M(n) = 0 \tag{10b,c}$$

(e.g., if each span is uniformly loaded with an intensity equal to q_o, $M_o = q_o \ell^2/2$). From Eq. A-1.9a, the particular solution is seen to be $M^p(r) = -M_o/6$; thus, for a total solution one must add a homogeneous solution with $M^h(0) = M^h(n) = M_o/6$. Then, by use of Table A-3, the total solution can be written as follows:

$$M(r) = -\frac{M_o}{6}\left[1 - (-1)^r F(n,r)\right] \tag{11a}$$

$$F(n,r) = \begin{bmatrix} \dfrac{\cosh \lambda\left(\frac{n}{2} - r\right)}{\cosh \lambda \frac{n}{2}} & n - \text{even} \\[2ex] \dfrac{\sinh \lambda\left(\frac{n}{2} - r\right)}{\sinh \lambda \frac{n}{2}} & n - \text{odd} \end{bmatrix} \tag{11b}$$

$$\cosh \lambda = 2 \tag{11c}$$

In a similar fashion, for an impulse loading, the model and solution for the discrete Kernel function (see Eq. A-1.12 and Table A-3) is

$$(\Delta + 6)K(r,\alpha) = -\delta_r^\alpha \tag{12a}$$

$$K(0,\alpha) = K(n,\alpha) = 0 \tag{12b,c}$$

$$K(r,\alpha) = \frac{(-1)^{r-\alpha}}{\sinh \lambda}\left[\sinh \lambda\,(r-\alpha)\not{U}\,(r-\alpha) - \sinh \lambda\,(n-\alpha)\,\frac{\sinh \lambda\, r}{\sinh \lambda\, n}\right] \tag{13}$$

For a slightly different model, consider the equilibrium displacement of a cable with equally spaced loads. Reduction of Eq. 8a to the one dimensional case (i.e., $S_r = 0$, $R_s = H$, $a_r = \ell$) gives the following model for simple end supports:

(14a) $$\Delta_r w(r) = -\frac{\ell}{H} P(r)$$

(14b,c) $$w(0) = w(n) = 0$$

for the case of a fill load; i.e., $P(r) = \sigma w(r)$ the governing equation becomes :

(15a) $$(\Delta_r + 2\gamma) w(r) = 0$$

(15b) $$\gamma = \frac{\ell\sigma}{2H}$$

Thus, the equation is homogeneous and for zero end deflections or any other homogeneous boundary conditions the model is that of an eigen value problem so that nonzero solutions exist only for certain values of the parameter. The equation and general solution should then be written

(16a) $$(\Delta_r + 2\gamma_k) w(r) = 0 \qquad 0 < \gamma_k < 2$$

(16b) $$w(r) = A_k \sin \lambda_k r + B_k \cos \lambda_k r$$

(16c) $$\cos \lambda_k = 1 - \gamma_k$$

It is intersting to note the analogous problem for the continuously loaded cable

(17a) $$D_x^2 w(x) = -\frac{1}{H} q(x)$$

(17b) $$q(x) = \upsilon w(x)$$

(17c) $$(D_x^2 + \alpha_i^2) w(x) = 0$$

(17d) $$\alpha_i^2 = \frac{\upsilon}{H}$$

(17e) $$w(x) = \overset{*}{A}_i \sin \alpha_i x + \overset{*}{B}_i \cos \alpha_i x$$

These discrete and continuous models and their solutions for certain boundary conditions are fundamental to the use of finite and infinite Fourier series for the solutions of discrete and continuous models respectively. This subject is developed in some detail in Appendix III.

Closed Form Analytical Solutions to Partial Difference Equations. – The techniques for writing general solutions to partial difference equations are closely analogous to those used to solve partial differential equations and are reasonably well presented in standard references(*). (Unfortunately, however, most of the examples are for lower order equations than those so often encountered by the structural analyst). The student with an interst in doing original work in discrete field mechanics should study the references on formal solutions to partial difference equations in detail; however, as in the continuum, a large portion of the models with two or more independent variables or coordinates are tractable by a striaghtforward separation of variables approach and use of a sinusoidal series to express the functions of at least one of the coordinates. Thus the practitioner can write solutions for many multidimensional systems without reference to the functional forms required for a homogeneous solution.

The purpose of this section will be to briefly review some of the classical approaches to solving partial difference equations as an aid to following or duplicating the solutions for structural systems covered in subsequent lectures.

Two sample partial difference equations encountered by the structural systems analyst are the equilibrium displacement equation for the doubly threaded net (constant coefficient version of Eq. 8a) and the diagonal stress function for a latticed hyperbolic paraboloid(**) shown as Eqs. 18 and 19 respectively below

$$\left[\frac{R}{a}\,\Delta\!\!\!/_r + \frac{S}{b}\,\Delta\!\!\!/_s\right] w(r,s) = -\,P(r,s) \qquad (18)$$

$$(E_r E_s + 1)T(r,s) = C_o \qquad (19)$$

In general, homogeneous partial difference equations with constant coefficients can be written as polynominals in E_r and E_s operating on the dependent variable

$$\psi(E_r\,,\;E_s)f(r,s) = 0 \qquad (20)$$

(*) "Calculus of Finite Differences", Charles Jordan, Chelsea, New York, 1950, pp. 605-644.

"Finite Difference Equations", H. Levy and F. Lessman, Sir Isaac Pitman & Sons, London, 1959, pp. 239-269.

(**) "On the Statics of Latticed Shells", Donald L. Dean, Publications International Association for Bridge and Structural Engineering, Vol. 25, December 1965, p. 72.

One method of solution is to factor the operator polynomial into first order factors and write the homogeneous solution as a sum of the solutions for these operators. The solution for a first order operator is usually written symbolically by treating the operators with respect to one variable as a constant with respect to the function of the other variable. For example, rewrite the homogeneous part of Eq. 19

(21a) $$(E_r + E_s^{-1})T^h(r,s) = 0$$

Thus from the first principal for solving ordinary difference equations

(21b) $$T^h(r,s) = (-1)^r (E_s^{-1})^r C(s)$$

or

(21c) $$T^h(r,s) = (-1)^r C(s-r)$$

or reversing treatment with respect to r and s

(21d) $$T^h(r,s) = (-1)^s C'(r-s)$$

A second method for finding homogeneous solutions to partial difference equations is the Lagrange method in which, much as with ordinary difference equations (see Eq. A-1.2b), the solution is written in exponential form, i.e.,

(22a) $$f(r,s) = \sum_i C(\alpha_i, \beta_i)\alpha_i^r \beta_i^s$$

in which, from Eq. 20,

(22b) $$\psi(\alpha_i, \beta_i) = 0$$

As an illustration, we will again use the homogeneous portion of Eq. 19.

(23a) $$\alpha_i \beta_i + 1 = 0$$

(23b) $$T^h(r,s) = \sum_i A\left(\alpha_i, \frac{-1}{\alpha_i}\right)\alpha_i^r \left(\frac{-1}{\alpha_i}\right)^s$$

(23c) $$= (-1)^s \sum_i A(\alpha_i)\alpha_i^{r-s}$$

in which $A(\alpha_i)$ is a set of arbitrary constants so that the summation is an arbitrary function of (r – s); thus, as before,

$$T^h(r,s) = (-1)^s C(r-s) \qquad (23d)$$

A particular solution to partial difference equations can often be obtained thru use of symbolic inversion of the operators. For example, consider the general constant coefficient case below to have a algebraic polynomial load term.

$$\psi(E_r, E_s) f(r,s) = V(r,s) \qquad (24a)$$

$$f^p(r,s) = \frac{V(r,s)}{\psi(E_r, E_s)} = \frac{V(r,s)}{\psi(1+\Delta_r, 1+\Delta_s)} \qquad (24b,c)$$

$$f^p(r,s) = [a_0 + a_1\Delta_r + b_1\Delta_s + a_2\Delta_r^2 + b_2\Delta_s^2 + C_2\Delta_r\Delta_s + \ldots]\ V(r,s) \qquad (24d)$$

For the case of V(r,s) equal to a finite algebraic polynomial in r and s, Eq. 24d terminates to an explicit formula for the particular solution.

Finally, we will consider a separation of variables approach to the solution of Eq. 18; i.e.,

$$w^h(r,s) = \sum_k X_k(r)\ Y_k(s) \qquad (25a)$$

Substitution into the homogeneous portion of Eq. 18 and a one step separation gives

$$\frac{\Delta\!\!\!\nabla_r X(r)}{X(r)} = -\frac{Sa}{Rb}\frac{\Delta\!\!\!\nabla_s Y(s)}{Y(s)} = -2\gamma_k \qquad (25b)$$

from which, by use of Table A-2, it is seen that:

$$\begin{aligned} X_k(r) &= A_k \cos\lambda_k r + B_k \sin\lambda_k r \\ \cos\lambda_k &= 1-\gamma_k \quad 0<\gamma_k<2 \end{aligned} \qquad (25c)$$

$$\begin{aligned} Y_k(s) &= C_k \cosh\sigma_k s + D_k \sinh\sigma_k s \\ \cosh\sigma_k &= 1 + \frac{Rb}{Sa}\gamma_k \end{aligned} \qquad (25d)$$

All that remains is to expand the loading term into a similar series and, for a useful but restricted class of boundary conditions, the solution may be readily found. For example, for $w(0,s) = w(m,s) = 0$, the load and solution can be expressed as a single sine series as follows:

$$\begin{bmatrix} w(r,s) \\ \\ P(r,s) \end{bmatrix} = \sum_{k=1}^{m-1} \begin{bmatrix} W_k(s) \\ \\ P_k(s) \end{bmatrix} \sin \frac{k\pi r}{m} \tag{26a}$$

Substitution into Eq. 18 and matching coefficients gives

$$\left(\Delta_s - 2 \frac{Rb}{Sa} \gamma_k\right) W_k(s) = - \frac{b}{s} P_k(s) \tag{26b}$$

$$W_k(s) = C_k \cosh \sigma_k s + D_k \sinh \sigma_k s + W_k^p(s) \tag{26c}$$

in which C_k and D_k are determined so as to satisfy the boundary conditions on s (e.g., at $s = 0,n$).

As will be demonstrated in subsequent lectures, it is often more convenient to also expand $W_k(s)$ as a sinusoidal series in s so that the total solution for w(r, s) is in the form of a double series.

LECTURE III

USE OF FINITE SINUSOIDAL SERIES

The Four Natural Boundary Conditions – One of the most useful solution forms for solving a discrete model is the finite sinusoidal series. The two principal advantages of the series over the classical exponential solution form are generality as to loading term and simpler algebra. The most straightforward use of a series solution is in those cases for which each term of the series satisfies the model boundary condtions; thus, the series is both a particular and a total solution. There are four types of boundary conditions (16 possible two boundary permutations) that can be identically satisfied by a closed form finite sinusoidal series. These cases are discussed in detail in Appendix III, pp. 139 thru 146.

Expansion of Given Load Functions. – In using finite sinusoidal series to solve discrete models, it is usually considered sufficient to express the coefficients of the series for the dependent variable in terms of the coefficients of the series for the load. The designer may then investigate response to any desired loading by expanding the load in terms of the appropriate series and substituting the coefficients into the solution series. Formulas for carrying out such expansions are given in Appendix III as follows: sine series (case a-a) – Eqs. A-3.35; cosine series (case d-d) – Eqs. A-3.40; displaced sine series (case c-c) – Eqs. A-3.44; displaced cosine series (case b-b) – Eqs. A-3.45; and mixed series case a-d, a-b, b-c, – Eqs. A-3.46, A-3.47, A-3.48 respectively. Several specific examples are shown in the Appendix III; e.g., A-3.37, A-3.38, A-3.41 and A-3.42. Other useful load expansions were tabulated in one of the author's earlier papers(*). For a detailed example, consider the expansion of a constant into a displaced sine series which is antisymmetric with respect to r = –1/2 and r = m – 1/2; (see Eq. A-3.44), i.e.,

$$1 = \sum_{k=1}^{m} A_k \sin \frac{k\pi}{m}\left(r + \frac{1}{2}\right) \tag{27a}$$

$$A_k = \frac{2\omega_k}{m} \Delta_r^{-1}\left[1 \cdot \sin \frac{k\pi}{m}\left(r + \frac{1}{2}\right)\right]\Bigg|_0^m \tag{27b}$$

From the table on p. 94 it is seen that the inverse delta operation gives

$$A_k = \frac{2\omega_k}{m}\left(\frac{-\cos \frac{k\pi r}{m}}{2 \sin \frac{k\pi}{2m}}\right)\Bigg|_0^m = \frac{\omega_k\,(1 - \cos k\pi)}{m \sin \frac{k\pi}{2m}} \tag{27c,d}$$

$$1 = \frac{2}{m} \sum_{k=1,3,\ldots}^{m} \frac{\omega_k}{\sin \frac{k\pi}{2m}} \sin \frac{k\pi}{m}\left(r + \frac{1}{2}\right) \tag{27e}$$

Then, by induction, one can write the expansion for other algebraic expressions as shown on pages 108 thru 109 for the sine series; i.e.,

$$1 - \frac{2}{m}\left(r + \frac{1}{2}\right) = \frac{2}{m} \sum_{k=2,4,\ldots}^{m} \frac{\omega_k}{\sin \frac{k\pi}{2m}} \sin \frac{k\pi}{m}\left(r + \frac{1}{2}\right) \tag{28}$$

(*) "Field Solutions for Two-Dimensional Frameworks", Donald L. Dean and Celina P. Ugarte, International Journal of Mechanical Sciences, Pergamon Press, Vol. 10, No. 4, April 1968, pp. 315-339.

$$(29)\quad \frac{m^2+1}{4} - \left(\frac{m-1}{2} - r\right)^2 = \frac{1}{m} \sum_{k=1,3,\ldots}^{m} \frac{\omega_k}{\sin^3\left(\frac{k\pi}{2m}\right)} \sin \frac{k\pi}{m}\left(r + \frac{1}{2}\right)$$

Transformation From Infinite to Finite Series. – A fundamental problem in the macro discrete field method is that of finding the finite series coefficients for a function given in the form of an infinite sinusoidal series evaluated at regularly spaced intervals; that is, the infinite series, which is valid for a continuous variable, is transformed to a finite series valid only for a corresponding discrete variable. This problem also is dealt with in Appendix III, pp. 115 thru 118 The expressions for the finite series coefficients appear in the form of a rapidly converging power series. (The rate of convergence increases with the number of intervals, m, in the range of interest). These series can be truncated after only a small number of terms, with satisfactory accuracy (e.g., one term for $I = 0$ only or three terms with $I = -1,(1),1$, etc.); or, in many cases, the series can be summed formally by use of standard reference formulas. One example is shown in Appendix III, pp. 116 and 117 and others will be carried out as steps in the solution of problems covered in subsequent lectures; however, for an additional detailed example, consider the transformation of the second order Kernel function below:

$$(30a)\qquad D_x^2 K(x,\xi) = -\delta(x-\xi)$$

$$(30b,c)\qquad K(0,\xi) = K(L,\xi) = 0$$

$$(30d)\qquad K(x,\xi) = \frac{2}{L}\sum_{i=1}^{\infty} \overset{*}{B}_i \sin\alpha_i\xi \sin\alpha_i x$$

$$(30e,f)\qquad \alpha_i = \frac{i\pi}{L} \qquad \overset{*}{B}_i = \frac{1}{\alpha_i^2}$$

It is required to find the coefficients of the discrete Kernel function, B_k, in the relation below:

$$(31a)\qquad \frac{2}{m}\sum_{k=1}^{m-1} B_k \sin\frac{k\pi a}{m} \sin\frac{k\pi r}{m} = \frac{2}{L}\sum_{i=1}^{\infty} \overset{*}{B}_i \sin\frac{i\pi a}{m} \sin\frac{i\pi r}{m}$$

Use of Eqs. A-3.54 and A-3.55 give

$$\frac{2}{m} B_k \sin \frac{k\pi a}{m} = \frac{2}{L} \sum_{I=-\infty}^{+\infty} \overset{*}{B}_{2Im+k} \sin \frac{\pi a}{m} (2Im + k) \qquad (31b)$$

$$B_k = \frac{m}{L}\left(\frac{L}{\pi}\right)^2 \sum_{I=-\infty}^{+\infty} (2Im + k)^{-2} \qquad (31c)$$

This summation is available from a standard reference(*)

$$B_k = \frac{L}{2m\gamma_k} \qquad \gamma_k = 1 - \cos \frac{k\pi}{m} \qquad (32a,b)$$

Satisfaction Of Unnatural Boundary Conditions. – Classical use of sinusoidal series to solve mathematical models is restricted to those cases for which each term of the series identically satisfies the model boundary condition, which must then be homogeneous. Such limitations are unacceptable in practice as the Fourier series solution has too many attractive features for it to be abandoned every time a model is encountered which has inhomogeneous boundary conditions or homogeneous boundary conditions differenct from those for which the series parameters are known in closed form. It is intuitively apparent that one need not accept the classical restrictions from the fact that the exact solution in say an algebraic form could be expanded into a sinusoidal series that would also be exact, at least on the interior; thus, all that is required is to devise a procedure for getting that series directly. The required procedure is to write the solution in the form of a series plus any function that does satisfy the boundary conditons for a total solution that satisfies the entire model, including boundary conditions. This boundary function can later be incorporated into the series for a solution in the form of a series only which may not satisfy the model on the boundaries. This latter series is identical to the above hypothetical series obtained by expanding the exact algebraic solution. As a first example of this procedure, consider the case of the prismatic continuous beam with an arbitrary interior loading and given end moments; i.e., the model (see Eq. 1) is

$$(\Delta_r + 6)M(r) = -6 \frac{B}{\ell} \Phi(r) \qquad (33a)$$

(*) Summation of Series, by L.B.W. Jolley, Dover Publications, New York, 1961, Eq. 822, p. 152.

(33b,c) $$M(0) = M^{\$} + M^{a/s} \quad , \quad M(n) = M^{\$} - M^{a/s}$$

The interior load and joint moment fields are written as finite sine series; i.e.,

(34) $$\Phi(r) = \sum_{k=1}^{n-1} \Phi_k \sin \frac{k\pi r}{n}$$

$$M(r) = M^{\$} + M^{a/s}\left(1 - 2\frac{r}{n}\right) + \sum_{k=1}^{n-1} A_k \sin \frac{k\pi r}{n} \qquad r = 0,(1),n$$
(35a,b)

Substitution into the governing equation gives:

$$\sum_{k=1}^{n-1} (6 - 2\gamma_k) A_k \sin \frac{k\pi r}{n} = -6\left[M^{\$} + M^{a/s}\left(1 - 2\frac{r}{n}\right)\right] - 6\frac{B}{\ell} \sum_{k=1}^{n-1} \Phi_k \sin \frac{k\pi r}{n}$$
(36a)

(36b) $$\gamma_k = 1 - \cos \frac{k\pi}{n}$$

The boundary function is expanded into a series as shown in Appendix III, p. 108 and A_k is determined by matching coefficients as follows:

(36c) $$A_k = \frac{-3}{3 - \gamma_k} \left(\frac{2}{n} \bar{M}_k \cot \frac{k\pi}{2n} + \frac{B}{\ell} \Phi_k\right)$$

(36d) $$\bar{M}_k = \begin{bmatrix} M^{\$} & \text{for} & k \text{ odd} \\ M^{a/s} & \text{for} & k \text{ even} \end{bmatrix}$$

Use of Eqs. 36c,d in Eq. 35 gives a solution valid for all joints. The series for the boundary function can be combined with A_k to give the following series valid over the range $r = 1,(1), n-1$

(37a) $$M(r) = -\sum_{k=1}^{n-1} \frac{3\frac{B}{\ell}\Phi_k + \frac{2}{n}\bar{M}_k \sin \lambda_k}{3 - \gamma_k} \sin \lambda_k r$$

(37b) $$\lambda_k = \frac{k\pi}{n}$$

As a second example of the procedure for using sinusoidal series to satisfy unnatural boundary conditions, consider the case of a prismatic continuous beam with an arbitrary symmetric interior loading and moment spring boundary conditions; i.e., the model is:

$$(\nabla_r + 6)M(r) = -\frac{6B}{\ell}\Phi^{\$}(r) \tag{38a}$$

$$\Phi^{\$}(r) = \Phi^{\$}(n-r) \tag{38b}$$

$$M(0) = -\kappa\theta(0) \qquad M(n) = \kappa\theta(n) \tag{39a,b}$$

or

$$\left(\Delta_r + 3 + \frac{6B}{\kappa\ell}\right)M(0) = -\frac{6B}{\ell}\Phi^{\$}(0) \tag{39c}$$

$$\left(\frac{6B}{\kappa\ell} + 3 - \nabla_r\right)M(n) = -\frac{6B}{\ell}\Phi^{\$}(n) \tag{39d}$$

The procedure for use of sinusoidal series to satisfy unnatural homogeneous boundary conditions; such as, Eqs. 39a,b requires two steps. First a solution is obtained for a problem with analogous inhomogeneous boundary conditions, such as the first example, Eq. 35a. The second step is to substitute the solution for inhomogeneous boundary conditions into the real boundary statement to find values for the pseudo inhomogeneous boundary quantities assumed in Step 1. For example, to solve the above model, Eqs. 38 and 39, one needs to substitute the solution to the first example, Eqs. 35 and 36, modified to account for symmetry, into Eq. 39c or 39d to find $M^{\$}$. That is,

$$M(r) = M^{\$} + \sum_{k=1,3,\dots}^{n-1} \frac{-3}{3-\gamma_k}\left(\frac{2}{n}M^{\$}\cot\frac{k\pi}{2n} + \frac{B}{\ell}\Phi_k\right)\sin\frac{k\pi r}{n} \tag{40}$$

Substitution of Eq. 40 into Eq. 39c yields the following equation for $M^{\$}$

$$\left[3 + \frac{6B}{\kappa\ell} - \frac{2}{n}\sum_k \frac{3\sin^2\frac{k\pi}{n}}{\gamma_k(3-\gamma_k)}\right]M^{\$} = -\frac{6B}{\ell}\Phi^{\$}(0) + \frac{B}{\ell}\sum_k \frac{3\Phi_k \sin\frac{k\pi}{n}}{3-\gamma_k} \tag{41a}$$

$$k = 1,(2),n-1 \tag{41b}$$

If one views Eq. 40 as a particular solution plus a homogeneous solution, i.e., $M(r) = M^p(r) + M^{\$}\bar{M}^h(r)$, then it can be seen that the second term on the right-hand side of Eq. 41a is simply $-M^p(1)$ and the series term in the brackets on the left-hand side of Eq. 41a equals $1 - \overline{M}^h(1)$. Thus, from the table of homogeneous solutions on p. 95 the second series can be formally summed and written as

$$(42a)\quad \frac{2}{n}\sum_{k=1,3,\ldots}^{n-1} \frac{3\sin^2\frac{k\pi}{n}}{\gamma_k(3-\gamma_k)} = \begin{bmatrix} 1 + \dfrac{\cosh\lambda\left(\frac{n}{2}-1\right)}{\cosh\frac{\lambda n}{2}} & n \quad \text{even} \\[2ex] 1 + \dfrac{\sinh\lambda\left(\frac{n}{2}-1\right)}{\sinh\frac{\lambda n}{2}} & n \quad \text{odd} \end{bmatrix}$$

$$(42b,c)\qquad \cosh\lambda = 2 \quad \text{or} \quad \lambda = \ln(2+\sqrt{3})$$

The solution of Eq. 41a for $M^{\$}$ and substitution into Eq. 40 completes the formula for $M(r)$ valid over the range $r = 0,(1),n$ or substitution of $M^{\$}$ for $\overline{M}_k$ and summing over $k = 1,(2), n-1$ in Eq. 37 yields a series only solution form valid over the range $r = 1,(1), n-1$. Of course, the series gives $M(0) = 0$ but the true value is $M(0) = M^{\$}$.

This completes the lecture on finite sinusoidal series but additional information on their use will be developed in subsequent lectures as this form of solution is employed for solving various discrete models.

LECTURE IV

DIFFERENCE MODELS AND SOLUTIONS FOR DISCRETE STRUCTURAL SYSTEMS

One-Dimensional Systems with Multiple Fields of Unknowns. – As a simple object problem for dealing with systems in the title category, consider the beam continuous over spring supports (see Fig. 2) whose span properties (excluding loads) are cyclic and symmetric with respect to the midpoint; i.e., in Eq. 5b, $\Delta_r\,(\ell/B)_r = 0$, $\alpha_r = \alpha'_r = \alpha$ and $\beta_r = \beta$ (for a prismatic beam $\alpha = 1/3$ and $\beta = 1/6$). The governing equation, derived via the "three moment" or flexibility approach is

$$\begin{bmatrix} \frac{\beta \ell}{B}(\Delta_r + 2\gamma) & \frac{1}{\ell}\Delta_r \\ \\ \frac{1}{\ell}\Delta_r & -\sigma \end{bmatrix} \begin{bmatrix} M(r) \\ \\ w(r) \end{bmatrix} = \begin{bmatrix} -\Phi(r) \\ \\ -P^e(r) \end{bmatrix} \tag{43}$$

in which the terms are as defined for Eqs. 3 and 5 and $\gamma = 1 + \alpha/\beta$ or one plus the reciprocal of the carry over factor. The governing equation derived via the "slope-deflection" or stiffness approach is:

$$\begin{bmatrix} (\Delta_r + 2\gamma) & -2\gamma\boxminus_r \\ \\ -2\gamma\boxminus_r & +2\gamma(\Delta_r - 2\bar{\sigma}) \end{bmatrix} \begin{bmatrix} \theta(r) \\ \\ \frac{1}{\ell}\, w(r) \end{bmatrix} = \frac{\ell}{bB} \begin{bmatrix} M^o(r) \\ \\ \ell P^o(r) \end{bmatrix} \tag{44}$$

in which $\bar{\sigma} = \ell^3\sigma/4\gamma bB$; $M^o(r)$ is the unbalanced moment at r if the joint deformations at $r-1$, r, $r+1$ are zero, that is, the applied moment minus fixed end moments or $(M^o(r) = M^a(r) - M^f(r) - M'^f(r-1)$; $P^o(r)$ is the unbalanced transverse force on r if deformations at $r-1$, r, $r+1$ are zero $(P^o(r) = P^a(r) + V^f(r) - V'^f(r-1)$ or applied force plus fixed end shears); and b is the opposite end stiffness coefficient or bB/ℓ equals the moment necessary to fix one end due to a unit rotation at the other end.

While the models Eqs. 43 and 44 have only 2 x 2 operator matrices, most of the discussion in this section will be aimed at solving larger systems as 2 x 2 matrices are often tractable by special "tricks" that are not applicable in general. Thus, it will be helpful to keep in mind the following two more general models:

$$\begin{bmatrix} a_{11} & a_{12} & \cdots & a_{1k} \\ a_{21} & a_{22} & \cdots & \\ \vdots & \vdots & & \\ a_{k1} & a_{k2} & \cdots & a_{kk} \end{bmatrix} \begin{bmatrix} x_1 \\ x_2 \\ \vdots \\ x_k \end{bmatrix} = \begin{bmatrix} p_1 \\ p_2 \\ \vdots \\ p_k \end{bmatrix} \tag{45}$$

$$(46) \qquad \begin{bmatrix} \alpha_{11} & \alpha_{12} \\ \alpha_{21} & a_{kk} \end{bmatrix} \begin{bmatrix} \bar{X}_1 \\ x_k \end{bmatrix} = \begin{bmatrix} \bar{P}_1 \\ p_k \end{bmatrix}$$

in which a_{ij} are a set of scalar linear operators (may be partial operators as well as ordinary operators and may have variable coefficients but must be commutative; i.e., $a_{k\ell}\, a_{ij} = a_{ij}\, a_{k\ell}$); x_j are scalar dependent variables and p_i are scalar load terms. The matrices in Eq. 46 are a subdivided version of Eq. 45; i.e., α_{11} is a k – 1 x k – 1 operator matrix, α_{12} is a k – 1 x 1 column matrix of operators, α_{21} is a 1 x k – 1 row matrix of operators, $\bar{X}_1$ is a k – 1 x 1 column matrix of dependent variables and $\bar{P}_1$ is a k – 1 x 1 column matrix of load terms.

One's first impulse for solving simultaneous equations is to successively eliminate unknows until there remains only one equation with one unknown (in which case the operator of the remaining equation would be the determinant of the operator matrix in the original set); and such a procedure would be relatively safe for Eq. 43 as one of the "operators" in a constant; i.e., one could eliminate w(r) by solving the second row for w(r) and substituting into first row to get a fourth order equation for M(r) in terms of Φ(r) and P^e(r). In general, however, the elimination of unknowns approach is not to be recommended as it may require more work than theoretically more straightforward procedures and, more importantly, one can very easily lose constants required to satisfy boundary conditions or introduce extraneous constants (or functions in the case of partial operators).

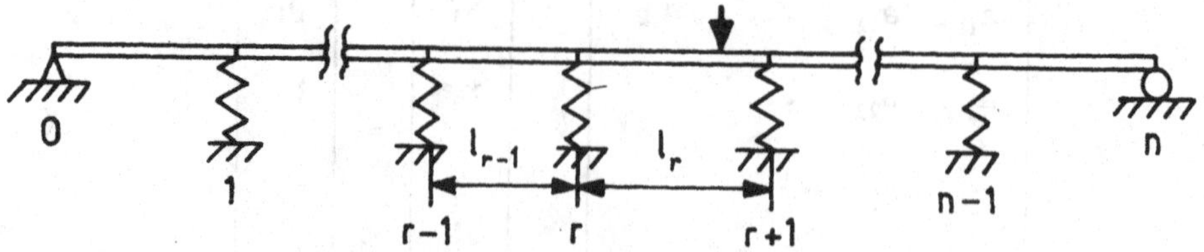

Fig. 2. Continuous Beam on Spring Supports

There are four established methods for solving linear models such as Eq. 45:

1) use of Cramer's rule;
2) use of a "stress" or potential function;
3) the Green's tensor approach; and
4) direct use of series.

Use of Cramer's rule to solve simultaneous equations with linear commutative operators can be indicated as follows :

$$|a_{ij}|\,[x_i] = [A_{ij}]\,[p_j] \tag{47}$$

in which $|a_{ij}|$ denotes the operator determinant and $[A_{ij}]$ is the adjoint matrix of $[a_{ij}]$ (i.e., the transpose of the matrix of cofactors). The order of $|a_{ij}|$, say m, determines the number of independent constants for the system. Note that in solving k equations for the x_i each with the same operator(*) the set of solutions will contain k x m apparent constants. One must then reduce the number of constants to m independent constants by back substituting the solutions into k – 1 rows of the original model and matching coefficients of like terms – a tedious process for k and/or m greater than 4.

Use of the "stress" function approach is usually to be preferred over use of Cramer's rule as the problem of introducing a large number of dependent constants is circumvented. The method consists of defining a new function, say $\psi(r)$, related to the components $x_i(r)$ so as to reduce k – 1 of the rows in Eq. 45 to identities and then to substitute into the kth row to get the governing equation for $\psi(r)$ – whose operator is $|a_{ij}|$, as in Cramer's rule, but here the operator is inverted only once. The subdivided from of Eq. 45, Eq. 46, was written as a guide in defining $\psi(r)$ so as to make the first k – 1 rows identities(**). The stress function definition and model can be indicated as follows:

$$\bar{X}_1(r) = [A_{ik}\,\psi(r)] + \alpha_{11}^{-1}\,\bar{P}_1 \qquad i = 1,(1),k-1 \tag{48a}$$

(*) Unless, as sometimes happens, the order can be reduced in some of the uncoupled equations by cancelling out like operator factors from both sides.

(**) Note that the approach is not unique as the definition of ψ depends upon the ordering of the equations. If all p_i except one are zero, the equations would normally be ordered so that p_k is the nonzero term.

(48b) $$x_k(r) = |\alpha_{11}|\,\psi(r)$$

(48c) $$|a_{ij}|\,\psi(r) = p_k - \alpha_{21}\,\alpha_{11}^{-1}\,\bar{P}_1$$

For an example of the use of Cramer's rule, consider Eq. 43. The uncoupled model is

(49a) $$\left[\nabla_r^2 + \frac{\beta\ell^3\sigma}{B}(\nabla_r + 2\gamma)\right]M(r) = -\sigma\ell^2\Phi(r) - \ell\nabla_r P^e(r)$$

(49b) $$\left[\nabla_r^2 + \frac{\beta\ell^3\sigma}{B}(\nabla_r + 2\gamma)\right]w(r) = -\ell\,\nabla_r\Phi(r) + \frac{\beta\ell^3}{B}(\nabla_r + 2\gamma)\,P^e(r)$$

Note that the homogeneous solutions are of the forms (see Eq. A-1.6)

$$\begin{bmatrix} M^h(r) \\ w^h(r) \end{bmatrix} = \begin{bmatrix} C_1 \\ C_1' \end{bmatrix} \cosh\eta r \cos\phi r + \begin{bmatrix} C_2 \\ C_2' \end{bmatrix} \cosh\eta r \sin\phi r$$

$$+ \begin{bmatrix} C_3 \\ C_3' \end{bmatrix} \sinh\eta r \cos\phi r + \begin{bmatrix} C_4 \\ C_4' \end{bmatrix} \sinh\eta r \sin\phi r$$

(49c)

in which the primed constants can be found in terms of the unprimed constants by back substitution in either row of Eq. 43.

As a first example of use of the "stress" function approach, consider Eq. 44. Selection of the stress function to make the first row an identity leads to the following definitions and scalar model:

(50a) $$\theta(r) = 2\gamma\Box_r\,\psi(r) + \frac{\ell}{bB}(\nabla_r + 2\gamma)^{-1}M^o(r)$$

(50b) $$\frac{1}{\ell}\,w(r) = (\nabla_r + 2\gamma)\,\psi(r)$$

$$\left[\nabla_r^2 + \frac{4\bar{\sigma}}{\gamma - 2}(\nabla_r + 2\gamma)\right]\psi(r) = -\frac{\ell}{b\gamma(\gamma-2)B}\left[\ell P^o(r) + \frac{2\gamma\Box_r}{\nabla_r + 2\gamma}M^o(r)\right]$$

(50c)

(Note that the above operator for the stiffness approach "stress" function, $\psi(r)$, is identical to the flexibility Cramer's rule operator as $\beta = 1/b\gamma\ (\gamma - 2)$).

As a second example of use of the "stress" function approach, consider Eq. 43, modified to make $a_{22} = 1$. Selecting the function to make the second row an identify gives:

$$w(r) = \frac{1}{\ell\sigma} \Delta_r \psi(r) + \frac{1}{\sigma} P^e(r) \qquad (51a)$$

$$M(r) = \psi(r) \qquad (51b)$$

$$\left[\Delta_r^2 + \frac{\beta\ell^3\sigma}{B}(\Delta_r + 2\gamma)\right] M(r) = \ell^2\sigma\Phi(r) - \ell\Delta_r P^e(r) \qquad (51c)$$

Thus, here, one of the variables, M(r), can serve as a "stress" function so that the uncoupled model is the same as would be found through elimination of unknowns.

A Green's tensor is the generalization of a Kernel or Green's function. For example, $K_{i\ell}(r,a)$ denotes (see Eq. 45) the solution for x_i due to $p_\ell = \delta_r^a$ and all other $p_i = 0$. That is, the set of Kernel functions comprising a Green's tensor is found by solving the system for a diagonal loading matrix of impulse functions. This could be used to determine the dependent variables due to any given loading as follows :

$$x_i(r) = \sum_{\alpha=1}^{n-1} [K_{ij}(r,\alpha)][P_j(\alpha)] \qquad (52)$$

the solution for the component Kernel functions can be indicated as follows:

$$K_{ij}(r,\alpha) = A_{ij}\psi(r,\alpha) \qquad (53a)$$

$$|a_{ij}|\psi(r,\alpha) = \delta_r^\alpha \qquad (53b)$$

In actual practice, the author has had difficulty in using Eqs. 53a and 53b to obtain Green's tensors at the stage where one evaluates the constants in the solution of Eq 53b to satisfy boundary conditions. It appears that each column of $[K_{ij}(r,\alpha)]$ (that is, for a given load vector consisting of only one impulse function) must be obtained as a separate problem for the determination of the arbitrary constants in $\psi(r,\alpha)$. In order words, one solves for a set of k distinct stress functions.

One could use series forms of solutions to solve any of the uncoupled models resulting from use of Cramer's rule, the stress function approaches or the Green's tensor approach; however, for many models, the step of first uncoupling the equations is unnecessary for a series approach. For example, to solve Eq. 43 with the boundary conditions $w\binom{0}{n} = M\binom{0}{n} = 0$, one may proceed as follows :

$$\text{(54a)} \qquad \begin{bmatrix} \Phi(r) \\ \\ P^e(r) \end{bmatrix} = \sum_{k=1}^{n-1} \begin{bmatrix} \Phi_k \\ \\ P_k \end{bmatrix} \sin \frac{k\pi r}{n}$$

$$\text{(54b)} \qquad \begin{bmatrix} M(r) \\ \\ w(r) \end{bmatrix} = \sum_{k=1}^{n-1} \begin{bmatrix} M_k \\ \\ W_k \end{bmatrix} \sin \frac{k\pi r}{n}$$

Substitution of Eqs. 54a and 54b into Eq. 43, matching coefficients and solving algebraically for the solution coefficients in terms of the load coefficients gives:

$$\text{(55a)} \qquad \begin{bmatrix} M_k \\ \\ W_k \end{bmatrix} = (1 + \zeta_k)^{-1} \begin{bmatrix} -\sigma\left(\frac{\ell}{2\gamma_k}\right)^2 & \frac{\ell}{2\gamma_k} \\ \\ \frac{\ell}{2\gamma_k} & \frac{1}{\sigma}\zeta_k \end{bmatrix} \begin{bmatrix} \Phi_k \\ \\ P_k \end{bmatrix}$$

in which

$$\text{(55b)} \qquad \gamma_k = 1 - \cos \frac{k\pi}{n}$$

$$\text{(55c)} \qquad \zeta_k = \frac{\beta\ell^3\sigma}{2B}\left(\frac{\gamma - \gamma_k}{\gamma_k^2}\right)$$

Analysis of the same physical system by use of the stiffness equations, Eq. 44, requires somewhat more care. In this case, the boundary statement portion of the model is:

$$\text{(56a)} \qquad \begin{bmatrix} \Delta_r + \gamma & -\gamma\Delta_r \\ \\ 0 & 1 \end{bmatrix} \begin{bmatrix} \theta(0) \\ \\ \frac{1}{\ell} w(0) \end{bmatrix} = \frac{\ell}{bB} \begin{bmatrix} \frac{1}{2}\bar{M}^o(0) \\ \\ 0 \end{bmatrix}$$

$$\text{(56b)} \qquad \begin{bmatrix} \gamma - \nabla_r & -\gamma\nabla_r \\ \\ 0 & 1 \end{bmatrix} \begin{bmatrix} \theta(n) \\ \\ \frac{1}{\ell} w(n) \end{bmatrix} = \frac{\ell}{bB} \begin{bmatrix} \frac{1}{2}\bar{M}^o(n) \\ \\ 0 \end{bmatrix}$$

in which the weighted quantities $\bar{M}^o(0)$ and $\bar{M}^o(n)$ are needed to work with a cosine series which is symmetric about its terminal boundary nodes(*). The solution for the series that satisfies Eq. 44 and the boundary conditions Eq. 56a and 56b simultaneously is as follows:

$$\begin{bmatrix} P^o(r) \\ w(r) \end{bmatrix} = \sum_{k=1}^{n-1} \begin{bmatrix} P_k \\ W_k \end{bmatrix} \sin\frac{k\pi r}{n} \tag{57a}$$

$$\begin{bmatrix} \bar{M}^o(r) \\ \theta(r) \end{bmatrix} = \sum_{k=0}^{n} \begin{bmatrix} \bar{M}^o_k \\ \Phi_k \end{bmatrix} \cos\frac{k\pi r}{n} \tag{57b}$$

$$\bar{M}^o_k = \frac{2\omega_k}{n} \sum_{r=0}^{n} \omega_r \bar{M}^o(r) \cos\frac{k\pi r}{n} \tag{57c}$$

$$= \frac{2\omega_k}{n} \sum_{r=0}^{n} M^o(r) \cos\frac{k\pi r}{n} \tag{57d}$$

Then substitution and matching coefficients as before gives

$$\begin{bmatrix} \theta_k \\ \frac{1}{\ell} W_k \end{bmatrix} = \frac{\ell}{2bBD_k} \begin{bmatrix} 2(\bar{\sigma} + \gamma_k) & \sin\lambda_k \\ \sin\lambda_k & \dfrac{\gamma - \gamma_k}{\gamma} \end{bmatrix} \begin{bmatrix} \bar{M}^o_k \\ \ell P_k \end{bmatrix} \tag{58a}$$

$$k = 1, (1), n-1$$

(*) There are three "tricks" in using the cosine series to solve discrete models that distinguish it from say the sine series: 1) the orthogonality is with respect to a weighting function (see Eq. A-3.40c); 2) the normalization factor contains a weighted function of the index number (see Eq. A-3.40e); and 3) the load term must be replaced by a weighted function equal to the given load function in the interior and equal to twice the given load function on the two boundary nodes.

in which $\lambda_k = \frac{k\pi}{n}$, $D_k = 2(\gamma - \gamma_k)(\bar{\sigma} + \gamma_k) - \gamma \sin^2\lambda_k$

(58d,e) $$\bar{\underline{\theta}}_0 = \frac{\ell}{2\gamma bB}\bar{\underline{M}}_0^o \; ; \quad \bar{\underline{\theta}}_n = \frac{\ell}{2(\gamma-2)bB}\bar{\underline{M}}_n^o$$

Two-Dimensional Systems with Single Field of Unknowns. – As an example of the title problem, consider the double layer flexural framework shown in Fig. 3. It is proportioned so that member shortenings is negligible throughout and the vertical web members have negligible flexural action. The model (which is identical to that for the doubly treaded net) can be derived (*) as follows:

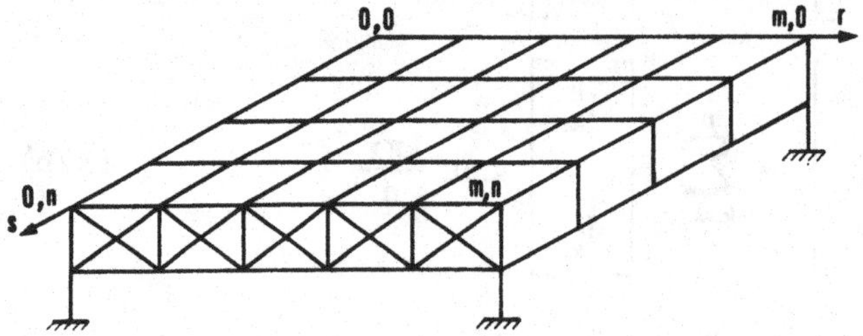

Fig. 3. Shear Grid with Stiff End Boundary Frames

(59a) $$\nabla_r V(r,s) + \nabla_s \bar{V}(r,s) = - P(r,s)$$

(59b,c) $$V(r,s) = K\Delta_r w(r,s), \; \bar{V}(r,s) = \bar{K}\Delta_s w(r,s)$$

(59d) $$(K \Delta\!\!\!\!\nabla_r + \bar{K} \Delta\!\!\!\!\nabla_s) w(r,s) = - P(r,s)$$

For the case of free side boundaries and very stiff end boundary frames, the boundary conditions are: $w(r, {}^0_n) = 0$, $\nabla_r w(0,s) = \Delta_r w(m,s) = 0$. Inspection of these boundary statements suggests the use of a sine series function of s (see case a-a p. 107 in Appendix III) and a displaced cosine series function of r (see case b-b p. 112) with m replaced by m + 1).

(*) Field Solutions for Shear Grids", Donald L. Dean, Journal of the Structural Division, ASCE Proceedings Paper 8594, Vol. 97, No. ST12, December 1971, p. 2848.

$$\begin{bmatrix} P(r,s) \\ \\ w(r,s) \end{bmatrix} \sum_{k=0}^{m} \sum_{\ell=1}^{n-1} \begin{bmatrix} P_{k\ell} \\ \\ W_{k\ell} \end{bmatrix} \cos \frac{k\pi}{m+1}\left(r + \frac{1}{2}\right) \sin \frac{\ell \pi s}{n} \tag{60}$$

Substitution of Eq. 60 into Eq. 59d and matching coefficients gives the following general solution:

$$W_{k\ell} = \frac{P_{k\ell}}{2K\gamma_k + 2\bar{K}\bar{\gamma}_\ell} \tag{61a}$$

$$\gamma_k = 1 - \cos \frac{k\pi}{m+1} \quad ; \quad \bar{\gamma}_\ell = 1 - \cos \frac{\ell n}{n} \tag{61b,c}$$

As an example of a specific loading consider the case of an impulse load; i.e., P(r,s) $= P_o \delta_r^\alpha \delta_s^\beta$

$$P_{k\ell} = \frac{4\omega_k}{mn} \sum_{r=0}^{m} \sum_{s=1}^{n-1} P(r,s) \cos \frac{k\pi}{m+1}\left(r + \frac{1}{2}\right) \sin \frac{\ell\pi s}{n} \tag{62a}$$

$$\omega_k = 1 - \frac{1}{2}\delta_k^0 \tag{62b}$$

$$= \frac{4\omega_k}{mn} \cos \frac{k\pi}{m+1}\left(a + \frac{1}{2}\right) \sin \frac{\ell\pi\beta}{n} \tag{62c}$$

Multi-Dimensional System with Multiple Fields of Unknowns. – One of the simpler structural systems in the title category is that of a plane grid loaded out-of-plane and having member properties and/or connection details such that the effects of torsion are negligible. For a flexibility approach to the difference model, the unknown joint quantities are the moments in the members parallel to the r axis M(r,s), the moments in the members parallel to the s axis, $\bar{M}$(r,s), and the out-of-plane displacements, w(r,s). The three partial difference equations are written by expressing rotational compatibility in the r direction, rotational compatibility in the s direction and out-of-plane force equilibrium. The model is (note similarity to the analogous one-dimensional model, Eq. 4.3):

$$(63)\qquad \begin{bmatrix} \frac{\beta a}{B}(\nabla_r + 2\gamma) & 0 & \frac{1}{a}\nabla_r \\ 0 & \frac{\bar{\beta}\bar{a}}{\bar{B}}(\nabla_s + 2\bar{\gamma}) & \frac{1}{\bar{a}}\nabla_s \\ \frac{1}{a}\nabla_r & \frac{1}{\bar{a}}\nabla_s & 0 \end{bmatrix} \begin{bmatrix} M(r,s) \\ \bar{M}(r,s) \\ w(r,s) \end{bmatrix} = - \begin{bmatrix} \Phi(r,s) \\ \overset{*}{\Phi}(r,s) \\ P^e(r,s) \end{bmatrix}$$

in which $\Phi(r,s)$ are the rotational discontinuities in the released r beams, $\overset{*}{\Phi}(r,s)$ are the rotational discontinuities in the released s beams, and $P^e\,(r,s)$ is the equivalent applied joint load including simple reactions due to midspan loads in the four spans joined at r,s. The operator determinant is fourth order with respect to both r and s; i.e.,

$$(64)\qquad |a_{ij}| = -\frac{\bar{\beta}\bar{a}}{\bar{B}a^2}\nabla_r^2(\nabla_s + 2\bar{\gamma}) - \frac{\beta a}{B\bar{a}^2}\nabla_s^2(\nabla_r + 2\gamma)$$

Thus, for a rectangular boundary(*) bounded by $r = 0$ and m and $s = 0$ and n, two conditions must be satisfied along the entire boundary. The simplest problem is that of a grid simply supported along all four edges of the boundary; i.e.,

$$(65)\qquad M\left(\begin{smallmatrix}0\\m\end{smallmatrix}, s\right) = \bar{M}\left(r, \begin{smallmatrix}0\\n\end{smallmatrix}\right) = w\left(\begin{smallmatrix}0\\m\end{smallmatrix}, s\right) = w\left(r, \begin{smallmatrix}0\\n\end{smallmatrix}\right) = 0$$

In this case the solutions and load components can all be expressed in double sine series:

$$(66a)\qquad \begin{bmatrix} \Phi(r,s) \\ \overset{*}{\Phi}(r,s) \\ P^e(r,s) \end{bmatrix} = \sum_{k=1}^{m-1}\sum_{\ell=1}^{n-1} \begin{bmatrix} \Phi_{k\ell} \\ \overset{*}{\Phi}_{k\ell} \\ P_{k\ell} \end{bmatrix} \sin\frac{k\pi r}{m}\,\sin\frac{\ell\pi s}{n}$$

(*) The reference is to rectangular in the mathematical sense as there is no reason to impose conditions of orthogonality between the two sets of beams for a torsionless grid.

$$\begin{bmatrix} M(r,s) \\ \bar{M}(r,s) \\ w(r,s) \end{bmatrix} = \sum_{k=1}^{m-1} \sum_{\ell=1}^{n-1} \begin{bmatrix} M_{k\ell} \\ \bar{M}_{k\ell} \\ W_{k\ell} \end{bmatrix} \sin \frac{k\pi r}{m} \sin \frac{\ell\pi s}{n} \tag{66b}$$

$$\begin{bmatrix} M_{k\ell} \\ \bar{M}_{k\ell} \\ W_{k\ell} \end{bmatrix} = \begin{bmatrix} -\frac{2\beta a}{B}(\gamma - \gamma_k) & 0 & +\frac{1}{2}2\gamma_k \\ 0 & -\frac{2\bar{\beta}\bar{a}}{\bar{B}}(\bar{\gamma} - \bar{\gamma}_\ell) & +\frac{1}{a}2\bar{\gamma}_\ell \\ +\frac{1}{a}2\gamma_k & +\frac{1}{\bar{a}}2\bar{\gamma}_\ell & 0 \end{bmatrix}^{-1} \begin{bmatrix} \Phi_{k\ell} \\ \overset{*}{\Phi}_{k\ell} \\ P_{k\ell} \end{bmatrix} \tag{67a}$$

in which
$$\gamma_k = 1 - \cos \frac{k\pi}{m} \qquad \bar{\gamma}_\ell = 1 - \cos \frac{\ell\pi}{n} \tag{67b,c}$$

or for the special case of no midspan loads, $\Phi_{k\ell} = \overset{*}{\Phi}_{k\ell} = 0$, the solution is

$$\begin{bmatrix} M_{k\ell} \\ \bar{M}_{k\ell} \\ W_{k\ell} \end{bmatrix} = \begin{bmatrix} \frac{4\bar{\beta}\bar{a}}{\bar{B}a}\gamma_k(\bar{\gamma} - \bar{\gamma}_\ell) \\ \frac{4\beta a}{B\bar{a}}\bar{\gamma}_\ell(\gamma - \gamma_k) \\ \frac{4\beta\bar{\beta}a\bar{a}}{B\bar{B}}(\gamma - \gamma_k)(\bar{\gamma} - \bar{\gamma}_\ell) \end{bmatrix} \frac{P_{k\ell}}{D_{k\ell}} \tag{67d}$$

in which

$$D_{k\ell} = 8\frac{\beta a}{B\bar{a}^2}\bar{\gamma}_\ell^2(\gamma - \gamma_k) + 8\frac{\bar{\beta}\bar{a}}{\bar{B}a^2}\gamma_k^2(\bar{\gamma} - \bar{\gamma}_\ell) \tag{67e}$$

The above solution for a grid simply supported along the entire boundary can be routinely extended to cover the case of a grid with flexible side supports. The first step is to add a homogeneous solution for boundary deflection along say r = 0 and m. The homogeneous solutions for M(r,s) and $\bar{M}$(r,s) would still be written as in Eq. 66b but the displacements would be in the form (see Eq. 35a for one dimensional analog):

$$w^h(r,s) = \sum_{\ell=1}^{n-1} \left[W_\ell^{\$} + W_\ell^{a/s}\left(1 - \frac{2r}{m}\right) + \sum_{k=1}^{m-1} \bar{W}_{k\ell} A_{k\ell} \sin\frac{k\pi r}{m} \right] \sin\frac{\ell\pi s}{n} \tag{68a}$$

$$\bar{W}_{k\ell} = \begin{bmatrix} W_\ell^{\$} & k \quad \text{odd} \\ & \\ W^{a/s} & k \quad \text{even} \end{bmatrix} \tag{68b}$$

Substitution of the homogeneous series into the homogeneous form of Eq. 63, expansion of the algebraic terms as a series and matching coefficients gives the homogeneous solutions in terms of $W_\ell^{\$}$ and $W_\ell^{a/s}$. These coefficients are then determined through analysis of the boundary beams (which may be loaded and have properties different from the interior beams) accounting for the shear at the ends of the first interior grid element spans; e.g., V(0,s) = 1/aM(1,s).

LECTURE V

SUMMATION MODELS AND SOLUTIONS FOR DISCRETE STRUCTURAL SYSTEMS

One-Dimensional Systems. – The goal of this section is the introduction of the macro discrete field metjod–the major alternative to the micro or difference equation method. As a first object problem, consider the case of a prismatic beam continuous over spring supports, a problem introduced earlier, Eq. 4, and solved by the micro approach in the previous lecture, Eq. 55. The unknowns selected for a macro flexibility approach are the interior reactions, R(r), for r = 1,(1),m – 1. The complete deflection field (including between node deflections) and compatibility relations are given by Eqs.4a and 4b. The solution will be by Fourier series, infinite series for the continuous fields and finite series for the node quantities. The Kernel function is obtained by a beam analysis; i.e.,

$$D_x^4 K(x,\xi) = \frac{1}{B}\,\delta(x-\xi) \tag{69a}$$

$$K\binom{0}{L},\xi\Big) = D_x^2 K\binom{0}{L},\xi\Big) = 0 \tag{69b,c}$$

$$K(x,\xi) = \frac{2}{L}\sum_{i=1}^{\infty} \overset{*}{B}_i \sin\alpha_i\xi \sin\alpha_i x \tag{69d}$$

$$\alpha_i = \frac{i\pi}{L} \quad ; \quad \overset{*}{B}_i = \frac{1}{B\alpha_i^4} \tag{69e,f}$$

The beam deflection due to applied load only, q(x); (i.e., R(r) = 0) is found as follows:

$$q(x) = \sum_{i=1}^{\infty} \overset{*}{q}_i \sin\alpha_i x \tag{70a}$$

$$w^o(x) = \int_0^L K(x,\xi)q(\xi)d\xi = \sum_{i=1}^{\infty} \overset{*}{q}_i \overset{*}{B}_i \sin\alpha_i x \tag{70b,c}$$

The reaction field is expressed as a finite series

$$R(r) = \sum_{k=1}^{m-1} R_k \sin\frac{k\pi r}{m} \tag{71a}$$

from which the inverse relation (see Eq. A-3.35d) is:

$$R_k = \frac{2}{m}\sum_{r=1}^{m-1} R(r)\sin\frac{k\pi r}{m} \tag{71b}$$

Note that substitution of the Kernel function series, Eq. 69d, into the summation term of Eq. 4a gives:

$$\sum_{\alpha=1}^{m-1} R(\alpha) \sum_{i=1}^{\infty} \frac{2}{L}\overset{*}{B}_i \sin\frac{i\pi\alpha}{m} \sin\alpha_i x$$

Reversal of the summation order and use of the formula for the finite series coefficients, Eq. 71b, reduces the term to(*)

$$\sum_{i=1}^{\infty} \frac{m}{L} R_i \overset{*}{B}_i \sin \alpha_i x$$

in which R_i is cyclic for indices outside the finite series range, i.e.,

(72a,b) $$R_k = - R_{2Im-k} = R_{2Im+k}$$

for I any integer. Substitution of this relation and Eq. 70b into Eq. 4a gives the following series for the continuous deflection field:

(73) $$w(x) = \sum_{i=1}^{\infty} \left[\overset{*}{q}_i - \frac{m}{L} R_i \right] \overset{*}{B}_i \sin \alpha_i x$$

To determine R(r), or R_k, use is made of the compatibility relation $1/\sigma\, R(r) = w(\frac{L}{m} r)$ for which $w(\frac{L}{m} r)$ is required in the form of a finite series. This transformation is made by usė of Eq. A-3.55c.

(74a) $$w\left(\frac{L}{m} r\right) = \sum_{k=1}^{m-1} \left[W_k^o - R_k B_k \right] \sin \frac{k\pi r}{m}$$

(74b) $$W_k^o = \sum_{I=-\infty}^{+\infty} (\overset{*}{q} \overset{*}{B})_{2Im+k}$$

(74c) $$B_k = \frac{m}{L} \sum_{I=-\infty}^{+\infty} \overset{*}{B}_{2Im+k}$$

The summation for B_k is available in closed form(**)

(74d,e) $$B_k = \frac{1}{12B} \left(\frac{L}{m}\right)^3 \frac{3-\gamma_k}{\gamma_k^2} \qquad \gamma_k = 1 - \cos \frac{k\pi}{m}$$

(*) As a mnemonic device note the form is that of a loading equal to the reactions smeared out over the element span length, $\ell = L/m$.

(**) "Macro Approach to Discrete Field Analysis", Donald L. Dean and Hota V.S. GangaRao, Journal of the Engineering Mechanics Division, ASCE Proceedings Paper 7463, Vol. 96, No. EM4, August 1970, pp. 377-394.

(**) Summation of Series by L.B.W. Jolley, Dover Publications, New York, 1961, Eq. 824, p. 154.

It should be noted in passing that B_k is interpretable as the coefficient of a discrete beam Kernel function

$$K\left(\frac{L}{m}r, \frac{L}{m}\alpha\right) = K'(r,\alpha) = \frac{2}{m}\sum_{k=1}^{m-1} B_k \sin\frac{k\pi\alpha}{m}\sin\frac{k\pi r}{m} \tag{75a}$$

and could be found alternatively via the micro approach by solving the difference equation for a discretely loaded beam (Eq. 3a with $\sigma = \Phi(r) = 0$ and $P^e(r) = \delta_r^\alpha$)

$$\Delta_r^2 K'(r,\alpha) = \frac{\ell^3}{6B}(\Delta_r + 6)\delta_r^\alpha \tag{75b}$$

$$K'\left(\frac{0}{m},\alpha\right) = 0 \tag{75c}$$

One can now obtain the desired expression for R_k by substituting Eq. 74a into the compatibility equation; i.e.

$$R_k = \frac{W_k^o}{\frac{1}{\sigma} + B_k} = \frac{\sigma W_k^o}{1 + \zeta_k} \tag{76a,b}$$

$$\zeta_k = \frac{\sigma\ell^3}{12B}\,\frac{3 - \gamma_k}{\gamma_k^2} \tag{76c}$$

As a first specific example, consider the case of a uniform load $q(x) = q_o$. By use of Eq. A-3.56c, it is seen that the coefficients of R(r) can be written in the form of a modification of the coefficients for a uniform reaction field of $q_o\ell$; i.e.,

$$R_k = \frac{6 - \gamma_k}{\frac{1}{\bar{\sigma}}\gamma_k^2 + 2(3 - \gamma_k)}(q_o\ell)\frac{2}{m}\cot\frac{k\pi}{2m} \tag{77a}$$

k = odd only

$$\bar{\sigma} = \frac{\sigma\ell^3}{24B} \tag{77b}$$

$$w\left(\frac{L}{m}r\right) = \frac{1}{\sigma}\sum_{k=1,3,\dots}^{m-1} R_k \sin\frac{k\pi r}{m} \tag{78a}$$

$$(78b)\qquad w(x) = \sum_{i=1,3,\ldots}^{\infty} \left[\frac{4q_o}{L\alpha_i} - \frac{m}{L} R_i \right] \frac{1}{B\alpha_i^4} \sin \alpha_i x$$

As a second example, consider the case of a harmonic continuous load, $q(x) = q_o \sin \pi x/L$. By use of Eqs. 70b and A-3.55c, it is seen that the unrestrained node deflections are also harmonic for the analogous value of k as related thru Eq. 72a; i.e., for $\overset{*}{q}_i = q_o \delta_i^1$

$$(79a)\qquad W_k^o = \frac{q_o}{B} \left(\frac{L}{\pi} \right)^4 \delta_k^1$$

$$(79b)\qquad w\left(\frac{L}{m} r \right) = \frac{q_o L^4 \sin \frac{\pi r}{m}}{\pi^4 B(1 + \zeta_1)}$$

The resulting continuous deflection field, however, it not harmonic as there are an infinite set of continuous deflection harmonics caused by the discrete harmonic reactions

$$(80a)\qquad R(r) = \frac{\sigma W_1^o}{1 + \zeta_1} \sin \frac{\pi r}{m}$$

i.e.,

$$(80b)\qquad i = 1,\ 2m - 1,\ 2m + 1,\ 4m - 1,\ 4m + 1,\ \ldots$$

Thus the continuous deflection field (see Eq. 73) is:

$$(81a)\qquad w(x) = \sum_{i=1}^{\infty} \left[q_o \delta_i^1 \mp \frac{m}{L} R_1 \delta_i^{2Im \pm 1} \right] \overset{*}{B}_i \sin \alpha_i x$$

or

$$(81b)\qquad w(x) = \left(q_o - \frac{m}{L} R_1 \right) \overset{*}{B}_1 \sin \frac{\pi x}{L} + \frac{mR_1}{L} \sum_{I=1}^{\infty} \left[\overset{*}{B}_{2Im-1} \sin \frac{\pi x}{L} (2Im - 1) - \overset{*}{B}_{2Im+1} \sin \frac{\pi x}{L} (2Im + 1) \right]$$

Two-Dimensional Systems. – The transition from one to two dimensional systems is easier with the macro field approach than with continuous systems. Here also we will select as a first problem one that has been previously solved via the micro field method. The structure to be analyzed is a simply supported torsionless grid. The approach will again be a flexibility analysis, but here there will be only one field of unknowns, the interactive node forces between the two sets of beams, R(r,s)—as constrasted with the three fields M(r,s), $\bar{M}$(r,s) and w(r,s) required for a micro field analysis (see Eq. 63). The complete deflection field for beams parallel to the r and x axes is

$$w(x,s) = w^{o}(x,s) - \sum_{\alpha=1}^{m-1} R(\alpha,s)K\left(x, \frac{L}{m}\alpha\right) \tag{82}$$

and the corresponding expression for the beams parallel to the s and y axes is

$$\bar{w}(r,y) = \bar{w}^{o}(r,y) + \sum_{\beta=1}^{n-1} R(r,\beta)\bar{K}\left(y, \frac{\bar{L}}{n}\beta\right) \tag{83}$$

in which w^{o} and $\bar{w}^{o}$ are the beam deflections due only to the applied loads (R(r,s) = 0) q(x,s) and $\bar{q}$(r,y) respectively; K and $\bar{K}$ are the Kernel functions for the beams parallel to the x and y axes respectively (see Eqs. 69); and the interactive forces, R(r,s), comprise the scalar field of unknowns to be determined. The necessary compatibility relations are given by

$$w\left(\frac{L}{m}r, s\right) = \bar{w}\left(r, \frac{\bar{L}}{n}s\right) \tag{84}$$

The analysis will be by Fourier series; i.e., double finite series for the node quantities and mixed infinite-finite series for the continuous beam deflection for the two fields of beams.

$$R(r,s) = \sum_{k=1}^{m-1}\sum_{\ell=1}^{n-1} R_{k\ell} \sin\frac{k\pi r}{m} \sin\frac{\ell\pi s}{n} \tag{85a}$$

$$R_{k\ell} = \frac{4}{mn}\sum_{r=1}^{m-1}\sum_{s=1}^{n-1} R(r,s) \sin\frac{k\pi r}{m} \sin\frac{\ell\pi s}{n} \tag{85b}$$

Then, as with the analogous one dimensional case, the continuous deflection fields, in terms of the unknown reaction coefficients, $R_{k\ell}$, are :

$$w(x,s) = \sum_{i=1}^{\infty}\sum_{\ell=1}^{n-1}\left[\overset{*}{q}_{i\ell} - \frac{m}{L}R_{i\ell}\right]\overset{*}{B}_{i} \sin\frac{\ell\pi s}{n} \sin\alpha_{i}x \tag{86a}$$

(86b) $$\bar{w}(r,y) = \sum_{k=1}^{m-1} \sum_{j=1}^{\infty} \left[\overset{*}{\bar{q}}_{kj} + \frac{n}{\bar{L}} R_{kj} \right] \overset{*}{\bar{B}}_j \sin \frac{k\pi r}{m} \sin \bar{\alpha}_j y$$

in which α_i and $\overset{*}{B}_i$ are as given by Eq. 69e,f

(86c,d) $$\bar{\alpha}_j = \frac{j\pi}{\bar{L}} \qquad \overset{*}{\bar{B}}_j = \frac{1}{\bar{B}\bar{\alpha}_j^4}$$

$R_{i\ell}$ is cyclic on i and with a period of 2m and R_{kj} is cyclic on j with a period of 2n and $\overset{*}{q}_{i\ell}$ and $\overset{*}{\bar{q}}_{kj}$ are coefficients of the applied beam loads; i.e.

(86e) $$q(x,s) = \sum_{i=1}^{\infty} \sum_{\ell=1}^{n-1} q_{i\ell} \sin \frac{\ell\pi s}{n} \sin \alpha_i x$$

(86f) $$\bar{q}(r,y) = \sum_{k=1}^{m-1} \sum_{j=1}^{\infty} \overset{*}{\bar{q}}_{kj} \sin \frac{k\pi r}{m} \sin \bar{\alpha}_j y$$

The solution for $R_{k\ell}$ results from transforming both sides of Eq. 84 to double finite series and matching coefficients; i.e.,

(87a) $$w\left(\frac{L}{m} r, s\right) = \sum_{k=1}^{m-1} \sum_{\ell=1}^{n-1} [W^o_{k\ell} - R_{k\ell} B_k] \sin \frac{k\pi r}{m} \sin \frac{\ell\pi s}{n}$$

(87b) $$\bar{w}\left(r, \frac{\bar{L}}{n} s\right) = \sum_{k=1}^{n-1} \sum_{\ell=1}^{n-1} [\bar{W}^o_{k\ell} + R_{k\ell} \bar{B}_\ell] \sin \frac{k\pi r}{m} \sin \frac{\ell\pi s}{n}$$

(87c) $$R_{k\ell} = \frac{W^o_{k\ell} - \bar{W}^o_{k\ell}}{B_k + \bar{B}_\ell}$$

in which B_k is as given by Eqs. 74d,e and

(87d,e) $$\bar{B}_\ell = \frac{1}{12\bar{B}} \left(\frac{\bar{L}}{n}\right)^3 \frac{3 - \bar{\gamma}_\ell}{\bar{\gamma}_\ell^2} \qquad \bar{\gamma}_\ell = 1 - \cos \frac{\ell\pi}{n}$$

(87f) $$W^o_{k\ell} = \sum_{I=-\infty}^{+\infty} (\overset{*}{q}_{2Im+k,\ell}) \overset{*}{B}_{2Im+k}$$

$$\overline{W}^{o}_{k\ell} = \sum_{J=-\infty}^{+\infty} (\overset{*}{\bar{q}}_{k,\,2Jn+\ell})\ \overset{*}{\bar{B}}_{2Jn+\ell} \tag{87g}$$

Note that for the case of a discrete loading say on the beams in the r direction

$$W^{o}_{k\ell} = P_{k\ell}\, B_{k} \tag{88}$$

in which $P_{k\ell}$ are the coefficients of the applied joint loading as shown in Eq. 66a.

Comparison with Solution to Difference Models. – Some systems can be analyzed by a micro discrete field approach but not by the macro approach while the reverse is true for other systems. The purpose of this section is to compare the solutions derived by the macro and micro approaches for certain lattices that have been analyzed by both methods.

For the first comparison, consider the problem of a beam continuous over spring supports. The governing equation for the micro flexibility analysis of the prismatic case is given by Eq. 3a while the nonprismatic case (a problem intractable by the macro approach) is covered by Eq. 43. The macro flexibility model is given by Eq. 4 or in series form by Eq. 73. Note that the macro model contains the continuous load term directly while the micro model shows only the joint effects of the midspan loads which must be determined by a separate analysis of the span elements. The sine series analysis of the micro flexibility model resulted in the following formula for the coefficients of the node deflection series (see Eq. 55)

$$W_{k} = (1 + \zeta_{k})^{-1}\left[\frac{\ell}{2\gamma_{k}}\,\Phi_{k} + \frac{1}{\sigma}\,\zeta_{k}P_{k}\right] \tag{89a}$$

while the same quantity was determined by the macro approach as (see Eq. 76)

$$W_{k} = \frac{1}{\sigma}\,R_{k} = \frac{W^{o}_{k}}{1 + \zeta_{k}} \tag{89b,c}$$

confidence that the results coincide for identical loads is gained by consideration of the case of a discrete loading with coefficients P_k. Then $\Phi_k = 0$ and $W^{o}_{k} = 1/\sigma\ \zeta_k P_k$ (see Eqs. 87 and 88).

The mciro solution has the advantage of validity for nonprismatic beams ($\gamma \neq 3$) and the macro solution has the advantage of giving the continuous deflection field in one step.(*)

For a second comparison, consider the closely related two-dimensional problem of a torsionaless grid. The micro flexibility model is given by Eq. 63 and the solution for the series coefficients of the node deflections is (see Eq. 67):

$$(90a)\qquad W_{k\ell} = \frac{1}{D_{k\ell}} \begin{bmatrix} \frac{4\bar{\beta}\bar{a}}{\bar{B}a}\,\gamma_k\,(\bar{\gamma}-\bar{\gamma}_\ell) \\ \\ \frac{4\beta a}{B\bar{a}}\,\bar{\gamma}_k\,(\gamma-\gamma_k) \\ \\ \frac{4\beta\bar{\beta}a\bar{a}}{B\bar{B}}\,(\gamma-\gamma_k)(\bar{\gamma}-\bar{\gamma}_\ell) \end{bmatrix}^T \begin{bmatrix} \Phi_{k\ell} \\ \\ \overset{*}{\Phi}_{k\ell} \\ \\ P_{k\ell} \end{bmatrix}$$

while the same quantity as determined by the macro approach is (see Eq. 87):

$$(90b)\qquad W_{k\ell} = \frac{\overset{o}{W}_{k\ell}\,\bar{B}_\ell + \overset{-o}{W}_{k\ell}\,B_k}{B_k + \bar{B}_\ell}$$

As before, confidence in the uniqueness of the solution is gained by comparing the two formulas for the case of joint loads only on a prismatic grid, i.e., $\beta = \bar{\beta} = 1/6$, $\gamma = \bar{\gamma} = 3$, $\Phi_{k\ell} = \overset{*}{\Phi}_{k\ell} = 0$, $ma = L$, $n\bar{a} = \bar{L}$, $\overset{-o}{W}_{k\ell} = 0$, $\overset{o}{W}_{k\ell} = B_k\,P_{k\ell}$ in which case both formulas give

$$(90c)\qquad W_{k\ell} = \frac{B_k\,\bar{B}_\ell}{B_k + \bar{B}_\ell}\,P_{k\ell}$$

Again, the advantages of the macro approach are simpler derivations and formulas (due to a single field of unknowns as opposed to three fields of unknowns required in the micro model) and a one step solution for continuous deflection fields while the advangage of the micro approach is the ability to deal with non-prismatic elements.

(*) This is an especially attractive feature for problems in which this system may be a component of a larger system with continuous interaction, of initially unknown magnitude, between the beam and say a plate component.

LECTURE VI

STRUCTURE WITH TIME DEPENDENT BEHAVIOR

Free and Forced Vibration of Structural Lattices. – Open form approaches to the analysis of structural systems are usually complicated to a significant degree by the transition from a statics to a dynamics analysis; however, for a field approach the addition of D'Alembert inertial forces or time dependent loads does not significantly complicate the analysis. As a working hypothesis, one could, over the protests of specialists in structural dynamics, characterize a trivial field problem as one whose solution is made much more complex by the addition of time dependent effects in the model. It is the goal of this section to illustrate use of the micro discrete field approach to the dynamic analysis of selected simple but nontrivial structural lattices.

As a first example, consider the shear grid with a lumped mass equal to ρ at each node. The model for a statical analysis is given by Eq. 59. Modification of that model for an undamped dynamic analysis of the same structure with uniform node masses is:

$$[\rho \, \underline{D}_t^2 - K\Delta\!\!\!/_r - \bar{K}\Delta\!\!\!/_s] \, w(r,s,t) = P(r,s,t) \qquad (91a)$$

$$w\left(r, {0 \atop n}, t\right) = \nabla_r \, w(0,s,t) = \Delta_r \, w(m,s,t) = 0 \qquad (91b)$$

The difference-differential model is separable as was the pure difference equation (Eq. 59) so that the solution can be written by expansion of w and P as double series in r and s with function of t as Euler coefficients; i.e. (see Eq. 60),

$$\begin{bmatrix} P(r,s,t) \\ \\ w(r,st) \end{bmatrix} = \sum_{k=0}^{m} \sum_{\ell=1}^{n-1} \begin{bmatrix} P_{k\ell}(t) \\ \\ W_{k\ell}(t) \end{bmatrix} \cos \lambda_k \left(r + \frac{1}{2} \right) \sin \bar{\lambda}_\ell \, s \qquad (92a)$$

$$\lambda_k = \frac{k\pi}{m+1} \;, \quad \bar{\lambda}_\ell = \frac{\ell\pi}{n} \qquad (92b,c)$$

Substitution into Eq. 91 and matching coefficients gives the following equation for the Euler functions:

(93a) $$(D_t^2 + \beta_{k\ell}^2)W_{k\ell}(t) = \frac{1}{\rho} P_{k\ell}(t)$$

(93b) $$\beta_{k\ell}^2 = \frac{2}{\rho}(K\gamma_k + \bar{K}\bar{\gamma}_\ell)$$

(93c,d) $$\gamma_k = 1 - \cos\lambda_k \qquad \bar{\gamma}_\ell = 1 - \cos\bar{\lambda}_\ell$$

Thus, the natural circular frequencies are given by $\beta_{k\ell}$ (i.e., period $= 2\pi/\beta_{k\ell}$) and the total solution can be easily written for any given variation of the load with respect to time. For example, if $P_{k\ell}(t)$ is harmonic, say $P_{k\ell}(t) = P_{k\ell}^o e^{i\sigma t}$ then the harmonic motions of the corresponding mode shapes ($\sigma \neq \beta_{k\ell}$) are:

(94) $$W_{k\ell}(t) = \frac{P_{k\ell}^o e^{i\sigma t}}{2(K\gamma_k + \bar{K}\bar{\gamma}_\ell) - \rho\sigma^2}$$

For a second specific example, consider the case of a unit impulse load applied at $t = t_o$; i.e., $P(r,s,t) = \delta_r^\alpha \delta_s^\beta \delta(t - t_o)$ or (see Eq. 62c)

(95a) $$P_{k\ell}(t) = P'_{k\ell}\delta(t - t_o)$$

(95b) $$P'_{k\ell} = \left(\frac{4\omega_k}{mn}\cos\lambda_k\left(\alpha + \frac{1}{2}\right)\sin\bar{\lambda}_\ell\beta\right)$$

with the structure initially at rest the solution is

$$w(r,s,t) = \sum_{k=0}^{m}\sum_{\ell=1}^{n-1}\frac{P'_{k\ell}}{\rho\beta_{k\ell}}[\sin\beta_{k\ell}(t - t_o)]U(t - t_o)\cos\lambda_k\left(r + \frac{1}{2}\right)\sin\bar{\lambda}_\ell s$$
(95c)

in which $U(t - t_o)$ is the Heaviside step function.

For a second problem, consider the case of shear frame, the one dimensional analog of the shear grid, which has lumped masses, ρ, at each floor level, and two types of viscous damping Newtonian bodies – an external damping, μ^e, that resists absolute node motion and an internal damping, μ^i , that resists relative motion between the nodes. The structural model is that of a m story single bay frame with stiff floor beams and the mathematical model is:

(96a) $$\nabla_r V(r,t) - (\rho D_t^2 + \mu^e D_t)w(r,t) = - P(r,t)$$

$$V(r,t) = (\mu^i D_t + K)\Delta_r w(r,t) \tag{96b}$$

$$[\rho D_t^2 + (\mu^e - \mu^i \nabla_r) D_t - K\nabla_r]\, w(r,t) = P(r,t) \tag{96c}$$

$$w(0,t) = \Delta_r w(m,t) = 0 \tag{96d,e}$$

For variations with respect to r, Eqs. 96d,e suggest a series of the type a-b (see Eq. A-3.47, Appendix III) with m replaced by m + 1; i.e.,

$$\begin{bmatrix} P(r,t) \\ \\ w(r,t) \end{bmatrix} = \sum_{k=1}^{m} \begin{bmatrix} P_k(t) \\ \\ W_k(t) \end{bmatrix} \sin \lambda_k r \tag{97a}$$

$$\lambda_k = \frac{2k-1}{2m+1}\pi \tag{97b}$$

Substitution into Eq. 96c and matching coefficients gives the following equation for the Euler functions, $W_k(t)$:

$$[\rho D_t^2 + (\mu^e + 2\gamma_k \mu^i) D_t + 2K\gamma_k] W_k(t) = P_k(t) \tag{98}$$

This equation can be solved routinely for a given load function and/or boundary conditions with respect to time. For example, for a given initial displacement configuration, $W_k(0) = W_k^o$ and $D_t W_k(0) = 0$ the solution is:

$$w(r,t) = \sum_{k=1}^{m} W_k^o (e)^{-\alpha_k t} \cos \beta_k t \times \sin \lambda_k r \tag{99a}$$

$$\alpha_k = \frac{1}{\rho}\left(\mu^i \gamma_k + \frac{1}{2}\mu^e\right) \tag{99b}$$

$$\beta_k^2 = \frac{2K}{\rho}\gamma_k - \alpha_k^2 \tag{99c}$$

$$\gamma_k = 1 - \cos \lambda_k \tag{99d}$$

For another specific example, consider the case of the system subjected to a unit impulse load;

$$P(r,t) = \delta_r^{\alpha} \delta(t - t_o) \tag{100a}$$

or

(100b)
$$P_k(t) = P_k'\delta(t - t_o)$$

(100c)
$$P_k' = \frac{4}{2m+1} \sin \frac{k\pi\alpha}{2m+1}$$

for which the solution is

$$w(r,t) = \sum_{k=1}^{m} \frac{P_k'}{\rho\beta_k} (e)^{-\alpha_k(t-t_o)} [\sin \beta_k (t - t_o)] U(t - t_o)] \sin \lambda_k r$$
(100d)

Systems with Viscoelastic Response .– In order to further illustrate the ease with which discrete field methods can be applied to systems with time dependent behavior, the goal of this section will be to demonstrate use of the macro discrete field approach for the analysis of a simple system with viscoelastic properties. This will be a quasi static analysis in that the force deformation changes occur so slowly with respect to the natural vibration periods that inertial terms can be omitted. The object problem will again be the continuous beam (see Fig. 4) but here it will be analyzed for interior supports that have viscoelastic responses instead of the simple Hookean response treated in Lecture V (see Eqs. 71 and 76). The compatibility relation used to analyse beams continuous over spring supports was

Fig. 4. Continuous Beam on Viscoelastic Supports

(101)
$$\frac{1}{\sigma} R(r) = w\left(\frac{L}{m} r\right)$$

whereas the relation required for general viscoelastic supports is of the form

(102a)
$$L_t R(r,t) = w\left(\frac{L}{m} r, t\right)$$

in which L_t is a differential operator whose form depends upon the particular combination of springs (Hookean bodies) and dashpots (Newtonian bodies) that comprise the typical viscoelastic body at each interior support point. The equations for node and continuous deflections are essentially unchanged from the spring support case except for the addition of a time parameter; i.e.,

$$w(x,t) = w^{o}(x,t) - \sum_{\alpha=1}^{m-1} R(\alpha,t)K\left(x, \frac{L}{m}\alpha\right) \tag{102b}$$

$$w\left(\frac{L}{m}r,t\right) = w^{o}\left(\frac{L}{m}r,t\right) - \sum_{\alpha=1}^{m-1} R(\alpha,t)K\left(\frac{L}{m}r, \frac{L}{m}\alpha\right) \tag{102c}$$

$$R(r,t) = \sum_{k=1}^{m-1} R_k(t) \sin\frac{k\pi r}{m} \tag{103a}$$

$$w(x,t) = \sum_{i=1}^{\infty} [\overset{*}{q}_i(t) - \frac{m}{L} R_i(t)] \overset{*}{B}_i \sin\alpha_i x \tag{103b}$$

$$w\left(\frac{L}{m}r,t\right) = \sum_{k=1}^{m-1} [W_k^{o}(t) - B_k R_k(t)] \sin\frac{k\pi r}{m} \tag{103c}$$

$$W_k^{o}(t) = \sum_{l=-\infty}^{+\infty} (\overset{*}{q}_{2lm+k}(t)) \overset{*}{B}_{2lm+k} \tag{103d}$$

Thus, by substituting Eqs. 103a and 103c into Eq. 102a, the governing equation for $R_k(t)$ is found to be

$$(L_t + B_k)R_k(t) = W_k^{o}(t) \tag{104}$$

As a first example, consider the case of a beam continuous over m – 1 interior Maxwell bodies and simply supported at the ends. That is, each interior support consists of a spring, σ, in series with a dashpot or viscous damping body, μ. Such bodies are often used to represent a support subject to creep. The general loading is considered to be applied instantaneously at time zero ($q(x,t) = q(x)$ $U(t - 0)$). Thus, for positive time, the unrestrained deflection fields are as given by Eqs. 70 and 74. For Maxwell bodies, the relation between force and displacement is

$$D_t X(t) = \left(\frac{1}{\sigma} D_t + \frac{1}{\mu}\right) F(t) \tag{105a}$$

or

$$L_t = \frac{\frac{1}{\sigma} D_t + \frac{1}{\mu}}{D_t} \tag{105b}$$

Substituting this operator into Eq. 104 and solving the first order differential equation for $R_k(t)$ gives.

(106a)
$$R_k(t) = \frac{\sigma W_k^o}{1 + \zeta_k} (e)^{(-t/\tau_k)}$$
in which

(106b)
$$\zeta_k = \frac{\sigma}{12B} \left(\frac{L}{m}\right)^3 \frac{3 - \gamma_k}{\gamma_k^2} \quad , \quad \gamma_k = 1 - \cos \frac{k\pi}{m}$$

and

(106c)
$$\tau_k = \frac{\mu}{\sigma} (1 + \zeta_k)$$

Note that the solution is in a form similar to that found for the spring support case (see Eq. 76b) but here it is apparent that the reactions decay, with characteristic time, τ_k , to zero and the displacements approach $w^o(x)$.

For a second example consider the Maxwell supports in the first example to be replaced by a three element body consisting of a spring, σ_s , in series with a Kelvin body, which consists of a spring, σ_p , in parallel with a dashpot, μ_p . Such bodies are often used to represent the behavior of a foundation on consolidating soil. For this type of support, the force-displacement relation is:

(107a)
$$X(t) = \left[\frac{1}{\sigma_s} + \frac{1}{\mu_p D_t + \sigma_p} \right] F(t)$$

(107b)
$$L_t = \frac{D_t + (\sigma_p + \sigma_s)/\mu_p}{\sigma_s [D_t + (\sigma_p/\mu_p)]}$$

and the solution is

(107c)
$$R_k(t) = \frac{\xi_k W_k^o}{\sigma_p + \xi_k} \left[\sigma_p + \xi_k (e)^{(-t/\tau_k)} \right]$$

(107d,e)
$$\xi_k = \frac{\sigma_s}{1 + \sigma_s B_k} \qquad \tau_k = \frac{\mu_k}{\sigma_p + \xi_k}$$

This completes the section on systems with viscoelastic response.

LECTURE VII

ANALYSIS OF MIXED DISCRETE–CONTINUOUS STRUCTURES

Ribbed Plates. – A flat plate reinforced by a series of parallel beams, sometimes called an orthotropic plate, is a highly efficient load carrying unit and is a good first example of a mixed discrete-continuous structure for analysis by a discrete field method (see Fig. 5). Many mixed discrete-continuous structures are multi-dimensional lattices with multi-mensional elements and so are intractable by the micro discrete field technique. With a single set of beams, however, the ribbed plate is a one dimensional lattice with two dimensional elements and so can be analyzed by either the micro or the macro field approach. A broad class of ribbed plate problems have been solved via the micro stiffness approach(*). If there are several simultaneous interactive forces between the ribs and the plates, the micro approach is attractive as the number of unknown node line displacements is independent of the number of interactive forces considered; for example, solutions that include torsional interaction between the rib and plate are now only available by the micro or difference equation approach. On the other hand, the necessity of getting an exact general boundary solution(**) for the plate strips between rib lines makes the algebra and derivation unattractive for systems with only one or two types of rib-plate interactive forces.

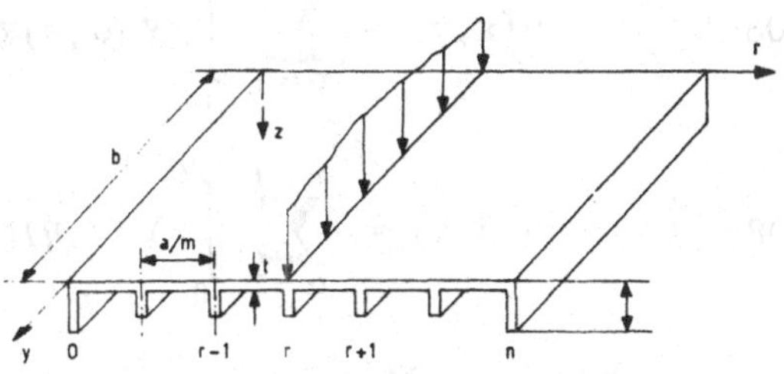

Fig. 5. Composite Membrane Model

Therefore, the principal object of this section will be to demonstrate use

(*) "Analysis of Ribbed Plates", Donald L. Dean and Cyrus Omid'varan, Journal of the Structural Division, ASCE Proceedings Paper 6474, Vol. 95, No. ST 3, March 1969, pp. 411-440.

(**) Extensive attempts to use highly accurate approximate boundary solutions for the sstrips were disappointing as a system with a large number of ribs is quite sensitive to even small inaccuracies in the strip boundary coefficients especially for ribs so stiff that the rib line displacements are small compared with the between rib displacements. This fact should alert one to the possibility of relatively meaningless results from an analysis by a standard open form finite element approach.

of the macro discrete field method for the composite membrane analysis of a simply supported ribbed plate. That is, the plate and rib components are proportioned and detailed so as to have negligible resistance to out-of-plane deformation; thus, the only interactive forces between the rib and plate are longitudinal forces parallel to the plane of the plate. Out-of-plane loads, N(r,y) are assumed to be applied only along the rib lines and the solution will be for the shear forces Y(r,y) along the top of the ribs which create a composite T-beam action between the ribs and the membrane.

The continuous membrane deflections are

$$u(x,y) = \sum_{\alpha=1}^{m-1} \int_0^b Y(\alpha,\eta) K^{uy}\left(x,y,\frac{a}{m}\alpha,\eta\right) d\eta \tag{108a}$$

$$v(x,y) = \sum_{\alpha=1}^{m-1} \int_0^b Y(\alpha,\eta) K^{vy}\left(x,y,\frac{a}{m}\alpha,\eta\right) d\eta \tag{108b}$$

in which K^{uy} and K^{vy} are the membrane Kernel functions given by Eqs. A-4.5 thru A-4.9 in Appendix IV. The in-plane and out-of-plane dispalcements at the top of the ribs are:

$$v_R(r,y) = \int_0^b [N(r,\eta)K^{vz}(y,\eta) + Y(r,\eta)K^{vy}(y,\eta)]\, d\eta \tag{109a}$$

$$w_R(r,y) = \int_0^b [N(r,\eta)K^{wz}(y,\eta) + Y(r,\eta)K^{wy}(y,\eta)]\, d\eta \tag{109b}$$

in which K^{vz}, K^{vy}, K^{wz}, and K^{wy} are the Kernel function elements of the Green's tensor due to loads on the rib tops (see Eqs. A-4.22 and A-4.23 in Appendix IV). The loads and interactive forces must be expressed in series form; i.e.,

$$N(r,y) = \sum_{j=1}^{\infty} \sum_{k=1}^{m-1} N_{kj} \sin\frac{k\pi r}{m} \sin \bar{\alpha}_j y \tag{110a}$$

$$Y(r,y) = \sum_{j=1}^{\infty} \sum_{k=1}^{m-1} Y_{kj} \sin\frac{k\pi r}{m} \cos \bar{\alpha}_j y \tag{110b}$$

$$\bar{\alpha}_j = \frac{j\pi}{b} \tag{110c}$$

Substitution of Eqs. 110 and the Kernel function solution into Eqs. 108 and 109 gives the following series for the membrane and rib top deflections:

$$u(x,y) = \sum_{i=1}^{\infty} \sum_{j=1}^{\infty} \frac{m}{a} Y_{ij} \overset{*}{A}_{ij} \cos \alpha_i x \sin \bar{\alpha}_j y \tag{111a}$$

$$v(x,y) = \sum_{i=1}^{\infty} \sum_{j=1}^{\infty} \frac{m}{a} Y_{ij} \overset{*}{B}_{ij} \sin \alpha_i x \cos \bar{\alpha}_j y \tag{111b}$$

$$\alpha_i = \frac{i\pi}{a}$$

$$v_R(r,y) = \sum_{j=1}^{\infty} \sum_{k=1}^{m-1} [N_{kj} \overset{*}{D}_j - Y_{kj} \overset{*}{B}_j] \sin \frac{k\pi r}{m} \cos \bar{\alpha}_j k \tag{112a}$$

$$w_R(r,y) = \sum_{j=1}^{\infty} \sum_{k=1}^{m-1} [N_{kj} \overset{*}{A}_j - Y_{kj} \overset{*}{D}_j] \sin \frac{k\pi r}{m} \sin \bar{\alpha}_j y \tag{112b}$$

in which Y_{ij} is cyclic on i with a period of 2m. The compatibility equation used to solve for the interactive force coefficients, Y_{kj} is

$$v\left(\frac{a}{m} r, y\right) = v_R(r,y) \tag{113}$$

for which the membrane rib line displacements are required in the form of an infinite-finite series. This can be obtained by transforming Eqs. 111 by use of Eq. A-3.55c in Appendix III or by direct use of the mixed continuous-discrete Kernel functions given by Eqs. A-4.12 thru A-4.16 in Appendix IV. By either method, the results are :

$$u\left(\frac{m}{a} r, y\right) = \sum_{j=1}^{\infty} \sum_{k=0}^{m} Y_{kj} A_{kj} \cos \frac{k\pi r}{m} \sin \bar{\alpha}_j y \tag{114a}$$

$$v\left(\frac{m}{a} r, y\right) = \sum_{j=1}^{\infty} \sum_{k=1}^{m-1} Y_{kj} B_{kj} \sin \frac{k\pi r}{m} \cos \bar{\alpha}_j y \tag{114b}$$

Matching coefficients between Eqs. 112a and 114b gives the solution for the interactive force coefficients as :

(115) $$Y_{kj} = \frac{\overset{*}{D}_j N_{kj}}{B_{kj} + \overset{*}{B}_j}$$

This comples the solution for the composite behavior of thin element ribbed plates. Any of the desired stress or deformation components can now be routinely computed by performing standard operations on the given deflection fields; for example, the rib moments are

(116) $$M_R(r,y) = - AE\rho^2 D_y^2 w_R(r,y)$$

This relatively straightforward technique for getting exact closed form solutions to ribbed plates can be easily extended for rather elegant solutions to more complicated systems. For example, by using a set of Kernel function solutions to the linear mathematical model for a cylindrical shell(*) one can write the exact solution for a cylindrical shell reinforced by evenly spaced longitudinal ribs. The solution to this problem is scheduled for publication in the near future(**).

Grid-Plate Structures. – A flat plate reinforced by two orthogonal sets of beams is an example of a two-dimensional lattice with two-dimensional elements and as such is an example of a system that cannot be analyzed rationally by a closed or open form micro approach due to the unavailability of general boundary solutions for the quadrilateral plate element bounded by successive beams in the two directions.

As the first problem in the title category consider the composite membrane analysis of a plate-stringer-diaphragm panel simply supported along its rectangular boundary. This structure is the same as the one considered in the previous section except here the composite ribs are themselves supported by a set of n – 1 interior diaphragms, flexural rigidity denoted by B^d, which have point contacts with he ribs or stringers but are not in direct contact with the top plate or membrane. The interactive forces between the stringers and diaphragms are denoted by R(r,s) – see grid analysis, Eqs. 82 thru 88 in Lecture V – ; i.e.,

---

(*)"Series Analysis of Cylindrical Shells – New Look at an Old Problem", R.R. Avent, Proceedings of the 13th Midwestern Mechanics Conference, University of Pittsburgh, August 13-15, 1973.

(**) "Field Analysis for Prefabricated Rbbed Cylindrical Shells", R. Richard Avent and Donald L. Dean, Proceedings IASS Symposium on Prefabricated Shells, Haifa, Israel, September 10-13, 1973.

$$R(r,s) = \sum_{k=1}^{m-1} \sum_{\ell=1}^{n-1} R_{k\ell} \sin \frac{k\pi r}{m} \sin \frac{\ell\pi s}{n}$$

Thus, for this problem, we need to solve for the coefficients of the interactive forces between the stringers and the membrane, Y_{kj} , and those of the interactive forces between the diaphragms and the stringers, $R_{k\ell}$, . The equations for the membrane deflections remain as shown in the previous section; i.e., Eqs. 108, 111, and 114. The stringer or rib deflections are changed to account for R(r,s) as follows:

$$v_R(r,y) = \sum_{j=1}^{\infty} \sum_{k=1}^{m-1} \left[\left(N_{kj} - \frac{n}{b} R_{kj} \right) \overset{*}{D}_j - \overset{*}{B}_j Y_{kj} \right] \sin \frac{k\pi r}{m} \cos \bar{\alpha}_j y \quad (117a)$$

$$w_R(r,y) = \sum_{j=1}^{\infty} \sum_{k=1}^{m-1} \left[\left(N_{kj} - \frac{n}{b} R_{kj} \right) \overset{*}{A}_j - \overset{*}{D}_j Y_{kj} \right] \sin \frac{k\pi r}{m} \sin \bar{\alpha}_j y \quad (117b)$$

in which R_{kj} is sinewise cyclic on j with a period of 2n. The node and continuous diaphragm deflections are:

$$w_D(x,s) = \sum_{i=1}^{\infty} \sum_{\ell=1}^{n-1} \frac{m}{a} R_{i\ell} \overset{*d}{B_i} \sin \alpha_i x \sin \frac{\ell\pi s}{n} \quad (118a)$$

in which $R_{i\ell}$ is sinewise cyclic on i with a period of 2m and $\overset{*d}{B_i}$ is as given by Eq. 69 e,f with B^d substituted for B.

$$w_D\left(\frac{a}{m} r, s\right) = \sum_{k=1}^{m-1} \sum_{\ell=1}^{n-1} R_k B_k^d \sin \frac{k\pi r}{m} \sin \frac{\ell\pi s}{n} \quad (118b)$$

in which B_k^d is as given by Eq. 74 with B replaced by B^d and L = a.

The compatibility equations are Eq. 113 and

$$w_R\left(r, \frac{b}{n} s\right) = w_D\left(\frac{a}{m} r, s\right) \quad (119)$$

in which w_R (r, b/n s) is required in the form of a double finite series. This can be found by use of Eq. A-3.55c in Appendix III as follows:

$$w_R\left(r, \frac{b}{n} s\right) = \sum_{k=1}^{m-1} \sum_{\ell=1}^{n-1} [W_{k\ell}^N - W_{k\ell}^Y - \bar{B}_\ell R_{k\ell}] \sin \frac{k\pi r}{m} \sin \frac{\ell\pi s}{n} \quad (120a)$$

in which $\bar{B}_\ell$ is given by Eq. 87d,e with $\bar{L} = b$ and B replaced by $\bar{B}$, and

(120b) $$W^N_{k\ell} = \sum_{J=-\infty}^{+\infty} N_{k,2Jn+\ell} \overset{*}{A}_{2Jn+\ell}$$

(120c) $$W^Y_{k\ell} = \sum_{J=-\infty}^{+\infty} \overset{*}{D}_{2Jn+\ell} Y_{k,2Jn+\ell}$$

Substitution of the appropriate series into the compatibility equation and matching coefficients gives;

(121a) $$(B_{kj} + \overset{*}{B}_j)Y_{kj} + \frac{n}{b}\overset{*}{D}_j R_{kj} = \overset{*}{D}_j N_{kj}$$

(121b) $$\sum_{J=-\infty}^{+\infty} \overset{*}{D}_{2Jn+\ell} Y_{k,2Jn+\ell} + (B_k + \bar{B}_\ell)R_{k\ell} = W^N_{k\ell}$$

The solution is:

(122a) $$Y_{kj} = \frac{\overset{*}{D}_j \left(N_{kj} - \frac{n}{b} R_{kj}\right)}{B_{kj} + \overset{*}{B}_j}$$

(122b) $$R_{k\ell} = \frac{W^N_{k\ell} - A^N_{k\ell}}{B_k + \bar{B}_\ell - \frac{n}{b}\bar{A}^R_{k\ell}}$$

(122c) $$A^N_{k\ell} = \sum_{J=-\infty}^{+\infty} \frac{(\overset{*}{D}_{2Jn+\ell})^2 N_{k,2Jn+\ell}}{B_{k,2Jn+\ell} + \overset{*}{B}_{2Jn+\ell}}$$

(122d) $$\bar{A}^R_{k\ell} = \sum_{J=-\infty}^{+\infty} \frac{(\overset{*}{D}_{2Jn+\ell})^2}{B_{k,2Jn+\ell} + \overset{*}{B}_{2Jn+\ell}}$$

These infinite series converge very rapidly and engineering accuracy is often obtained by use of the single term for $J = 0$. In such a case, one could truncate the series in Eq. 121b to only $\overset{*}{D}_\ell Y_{k\ell}$ for $\ell = 1,(1),n-1$ and solve Eq. 121a and b as simultaneous algebraic equations for $Y_{k\ell}$ and $R_{k\ell}$. The cyclic properties of R_{kj} could then be utilized in Eq. 121a to get good approximations for the higher harmonics of Y_{kj} . Such an approach is very useful as a cyclic algorithm for solving equations similar to Eq. 121 when each equation contains a summation with respect to one variable. Indeed, this is the type of model one must solve when both the diaphragms and stringers are in continuous contact with the plate. The solution to the grid-plate when both sets of beams are in continuous contact with the plate is suggested as a reader exercise.

LECTURE VIII

STRUCTURES WITH PLATE COMPONENTS

Flat Plate with In–Line Column Pattern– In the previous lectures the macro-discrete field method was introduced and applied to a variety of continuous beam problems. Mathematical solutions for continuous beams are not very exicting to the designer interested in the innovative application of complex structural system sand repeated use of such simple structures for illustrative purposes may fail to motivate the student for continued study of the subject. This would be unfortunate. Such simple illustrative problems were selected because their solutions by other techniques are well known to the audience. It was hoped this would facilitate assimilation of the new methods through their application to familiar systems. It will be the purpose of this lecture to demonstrate the ease with which the new methods can be used to solve more complex systems many of which are intractable by classical methods. The problems will be selected with a view toward the relative elegance of the solution and the heuristic value of the structural system rather than attempting to provide the most general possible set of design formulas.

The object problem of this section is the analysis of a special flat plate simply supported on its rectangular boundary and supported in the interior by a two dimensional set of in-line columns similar to the system shown in Fig. 6 except that the columns here are assumed to exert only concentrated reactive forces on the plate(*). The possibility of some column shortening can be easily included by modeling the column as an axial spring with constant σ. Thus, the structure is

---

(*) Extension to cover the case of moment as well as force interaction is routine and has been done as a student project at North Carolina State University.(**)

(**) "Analysis of a Flat Plate Continuous Over Column Supports", Dale N. Lee, Civil Engineering Master of Science Thesis, North Carolina State University, Raleigh, North Carolina, 1969.

simply the two dimensional analogue of the beam on spring supports which was dealt with in Lecture V (see Eqs. 73 and 76).

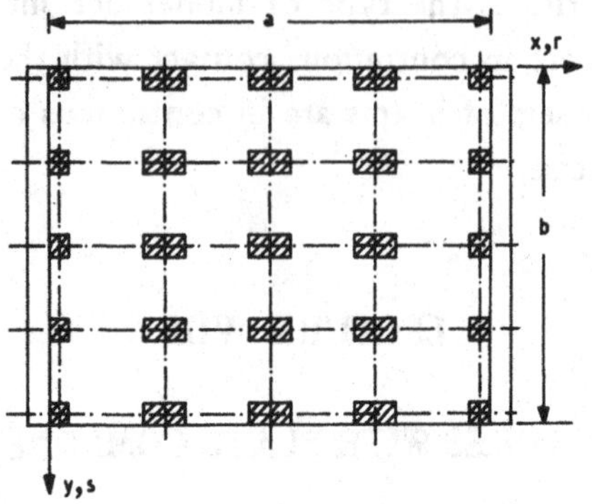

Fig. 6. Flat Plate with In-Line Columns

The deflection field is

$$w(x,y) = w^{o}(x,y) - \sum_{\alpha=1}^{m-1} \sum_{\beta=1}^{n-1} R(\alpha,\beta) K\left(x, y, \frac{a}{m}\alpha, \frac{b}{n}\beta\right) \tag{123}$$

in which R(r,s) is the unknown field of reactive forces

$$R(r,s) = \sum_{k=1}^{m-1} \sum_{\ell=1}^{n-1} R_{k\ell} \sin\frac{k\pi r}{m} \sin\frac{\ell\pi s}{n} \tag{124}$$

K is the Kernel function giving the flexural plate deflections at x,y due to a unit impulse load at ξ, η,

$$K(x,y,\xi,\eta) = \frac{4}{ab} \sum_{i=1}^{\infty} \sum_{j=1}^{\infty} C^{*}_{ij} \sin\alpha_i\xi \sin\bar{\alpha}_j\eta \sin\alpha_i x \sin\bar{\alpha}_j y \tag{125a}$$

$$\alpha_i = \frac{i\pi}{a}, \quad \bar{\alpha}_j = \frac{j\pi}{b}; \tag{125b,c}$$

$$\overset{*}{C}_{ij} = \frac{1}{D(\alpha_i^2 + \bar{\alpha}_j^2)^2} \tag{125d}$$

$$D = \frac{Et^2}{12(1-\mu^2)} \tag{125e}$$

and $w^o(x,y)$ = the deflection of the unrestrained plate; i.e., due to the applied loads $q(x,y)$ with $R(r,s) = 0$. The series for the complete plate deflection field is:

$$w(x,y) = \sum_{i=1}^{\infty}\sum_{j=1}^{\infty}\left[\overset{*}{q}_{ij} - \frac{mn}{ab} R_{ij}\right]\overset{*}{C}_{ij} \sin \alpha_i x \sin \bar{\alpha}_j y \tag{126}$$

in which R_{ij} is cyclic on i with a period of 2m and on j with a period of 2n. As in the beam problem, the compatibility relation is

$$\frac{1}{\sigma} R(r,s) = w\left(\frac{a}{m} r, \frac{b}{s} n\right) \tag{127}$$

Here, too, it is necessary to express the node deflections as a finite series:

$$w\left(\frac{a}{m} r, \frac{b}{n} s\right) = \sum_{k=1}^{m-1}\sum_{\ell=1}^{n-1}[W_{k\ell}^o - R_{k\ell}C_{k\ell}] \sin \frac{k\pi r}{m} \sin \frac{\ell\pi s}{n} \tag{128a}$$

$$W_{k\ell}^o = \sum_{I=-\infty}^{+\infty}\sum_{J=-\infty}^{+\infty}(\overset{*}{q}\overset{*}{C})_{2Im+k,\,2Jn+\ell} \tag{128b}$$

$$C_{k\ell} = \frac{mn}{ab}\sum_{I=-\infty}^{+\infty}\sum_{J=-\infty}^{+\infty}\overset{*}{C}_{2Im+k,\,2Jn+\ell} \tag{128c}$$

The author has not yet found an expression for the $C_{k\ell}$ double summation in elementary functions, but the same formulas used to get the beam summation gives the following alternate form of Eq. 128c(*)

$$C_{k\ell} = \frac{n}{b}\sum_{J=-\infty}^{+\infty}\bar{C}_{k,\,2Jn+\ell} \tag{129a}$$

(*) In programming, it is usually more convenient to use the double sum say for $I,J = -2,(1),+2$. For manual computations, it is often quite satisfactory to truncate the formula after the first term or $C_{k\ell} \simeq mn/ab\ \overset{*}{C}_{k\ell}$.

(129b) $$\bar{C}_{kj} = \frac{a}{4m D \bar{\alpha}_j^2 \bar{D}_{kj}} \left[\frac{\sinh \lambda_j}{\lambda_j} - \frac{1 - \cosh \lambda_j \cos \frac{k\pi}{m}}{\bar{D}_{kj}} \right]$$

(129c,d) $$\lambda_j = \frac{a}{m} \bar{\alpha}_j \quad ; \quad \bar{D}_{kj} = \cosh \lambda_j - \cos \frac{k\pi}{m}$$

The solution for $R_{k\ell}$ results from use of Eqs. 128a and 124 in Eq. 127; i.e.,

(130) $$R_{k\ell} = \frac{W^o_{k\ell}}{\frac{1}{\sigma} + C_{k\ell}}$$

As with the one dimensional problem, the special case of inflexible supports is covered by setting $\sigma = \infty$.

Flat Plate with Staggered Column Pattern. – The object problem of this section is the analysis of a flat plate similar to the one solved in the previous section except that here the columns are in a staggered pattern as shown in Fig. 7 and we provide for the possibility of differing spring constants for the offset columns; i.e., at r, s or $r + 1/2, s + 1/2$(*).

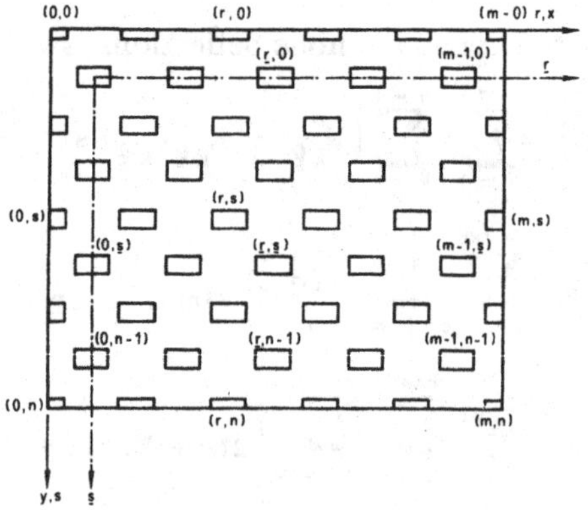

Fig. 7. Flat Plate with Staggered Columns

The continuous deflection field is:

$$w(x,y) = w^o(x,y) - \sum_{\alpha=1}^{m-1} \sum_{\beta=1}^{n-1} R(\alpha,\beta) K\left[x, y, \frac{a}{m}\alpha, \frac{b}{n}\beta \right] -$$

(*) The capability of analyzing systems of elements with alternating physical properties and/or geometry is important as such patterns usually offer increased structural efficiency as compared with in-line patterns.

$$- \sum_{\alpha=0}^{m-1} \sum_{\beta=0}^{n-1} \bar{R}(\alpha,\beta) K\left[x, y, \frac{a}{m}\left(\alpha + \frac{1}{2}\right), \frac{b}{n}\left(\beta + \frac{1}{2}\right)\right] \tag{131}$$

In which, w^o, R, and K are as defined for Eq. 123 and $\bar{R}(r,s)$ is the field of reactions due to the offset columns.

$$\bar{R}(r,s) = \sum_{k=1}^{m} \sum_{\ell=1}^{n} \bar{R}_{k\ell} \sin\frac{k\pi}{m}\left(r + \frac{1}{2}\right) \sin\frac{\ell\pi}{n}\left(s + \frac{1}{2}\right) \tag{132a}$$

$$\bar{R}_{k\ell} = \frac{4\omega_k \bar{\omega}_\ell}{mn} \sum_{r=0}^{m-1} \sum_{s=0}^{n-1} \bar{R}(r,s) \sin\frac{k\pi}{m}\left(r + \frac{1}{2}\right) \sin\frac{\ell\pi}{m}\left(s + \frac{1}{2}\right) \tag{132b}$$

$$\omega_k = 1 - \frac{1}{2}\delta_k^m \qquad \bar{\omega}_\ell = 1 - \frac{1}{2}\delta_\ell^n \tag{132c,d}$$

Substitution into Eq. 131 gives the following series for the deflection in terms of $R_{k\ell}$ and $\bar{R}_{k\ell}$

$$w(x,y) = \sum_{i=1}^{\infty} \sum_{j=1}^{\infty} \left[\overset{*}{q}_{ij} - \frac{mn}{ab} R_{ij} - \frac{mn}{ab} \frac{\bar{R}_{ij}}{\phi_i \bar{\phi}_j}\right] \overset{*}{C}_{ij} \sin\alpha_i x \sin\bar{\alpha}_j y \tag{133a}$$

in which $R_{k\ell}$ and $\bar{R}_{k\ell}$ are cyclic outside the normal range as follows:

$$R_{k\ell} = R_{2Im+k,\, 2Jn+\ell}\,; \quad \bar{R}_{k\ell} = (-1)^{I+J} \bar{R}_{2Im+k,\, 2Jn+\ell} \tag{133b,c}$$

for all integer values of I and J.

$$\phi_i = 1 - \frac{1}{2}\delta_i^{m(2I+1)} \qquad \bar{\phi}_j = 1 - \frac{1}{2}\delta_j^{n(2J+1)} \tag{133d,e}$$

The two compatibility equations required to find $R_{k\ell}$ and $\bar{R}_{k\ell}$ are

$$\frac{1}{\sigma} R(r,s) = w\left(\frac{a}{m} r, \frac{b}{n} s\right) \tag{134a}$$

$$\frac{1}{\bar{\sigma}} \bar{R}(r,s) = w\left[\frac{a}{m}\left(r + \frac{1}{2}\right), \frac{b}{n}\left(s + \frac{1}{2}\right)\right] \tag{134b}$$

Transformation of the right-hand sides into the appropriate double finite series in accordance with Eqs. A-3.55c and A-3.60b and matching coefficients yields:

(135a)
$$\frac{1}{\sigma} R_{k\ell} = W^{o}_{k\ell} - C_{k\ell} R_{k\ell} - \bar{C}_{k\ell} \bar{R}_{k\ell}$$
$$\frac{1}{\bar{\sigma}} \bar{R}_{k\ell} = \bar{W}^{o}_{k\ell} - \bar{C}_{k\ell} R_{k\ell} - C_{k\ell} \bar{R}_{k\ell}$$

in which $W^{o}_{k\ell}$ and $C_{k\ell}$ are as given by Eqs. 128b and 128c and $\bar{W}^{o}_{k\ell}$ and $\bar{C}_{k\ell}$ are the corresponding coefficients for transformation of the displaced sine series; i.e.,

(136a)
$$\bar{W}^{o}_{k\ell} = \omega_k \bar{\omega}_\ell \sum_{I=-\infty}^{+\infty} \sum_{J=-\infty}^{+\infty} (-1)^{I+J} (\overset{*}{q} \overset{*}{C})_{2Im+k,\ 2Jn+\ell}$$

(136b)
$$\bar{C}_{k\ell} = \omega_k \bar{\omega}_\ell \frac{mn}{ab} \sum_{I=-\infty}^{+\infty} \sum_{J=-\infty}^{+\infty} (-1)^{I+J} \overset{*}{C}_{2Im+k,\ 2Jn+\ell}$$

for k = m and/or ℓ = n the corresponding deflection mode is such as to give zero deflection at the full nodes; i.e., $R_{kn} = R_{m\ell} = 0$

(137a,b)
$$\bar{R}_{m\ell} = \frac{\bar{W}^{o}_{m\ell}}{\frac{1}{\bar{\sigma}} + C_{m\ell}} \quad ; \quad \bar{R}_{kn} = \frac{\bar{W}^{o}_{kn}}{\frac{1}{\bar{\sigma}} + C_{kn}}$$

For other values of $R_{k\ell}$ and $\bar{R}_{k\ell}$ the solution is

(138)
$$\begin{bmatrix} R_{k\ell} \\ \bar{R}_{k\ell} \end{bmatrix} = \begin{bmatrix} \frac{1}{\sigma} + C_{k\ell} & \bar{C}_{k\ell} \\ \bar{C}_{k\ell} & \frac{1}{\bar{\sigma}} + C_{k\ell} \end{bmatrix}^{-1} \begin{bmatrix} W^{o}_{k\ell} \\ \bar{W}^{o}_{k\ell} \end{bmatrix}$$

Discrete Core Sandwich Systems. – The standard solid core sandwich panel is well proven as a light-weight relatively rigid structural unit. In many instances for functional reasons and for efficient use of material it is advisable to replace the solid core of weak meterial by a discrete or latticed core of higher strength material. The resulting discrete core sandwich panel is well suited for analysis via discrete field methods. For a specific example, consider the panel shown in Fig. 8. The system is symmetric with respect to its middle plane and is loaded antisymmetrically; thus, the top and bottom plates have the same out-of-plane deflections. The plates are simply supported and are proportioned so that in-plane displacements are negligible as compared with out-of-plane displacements. Again the macro field approach must be used and the unknowns are the vertical interactive forces between the truss work

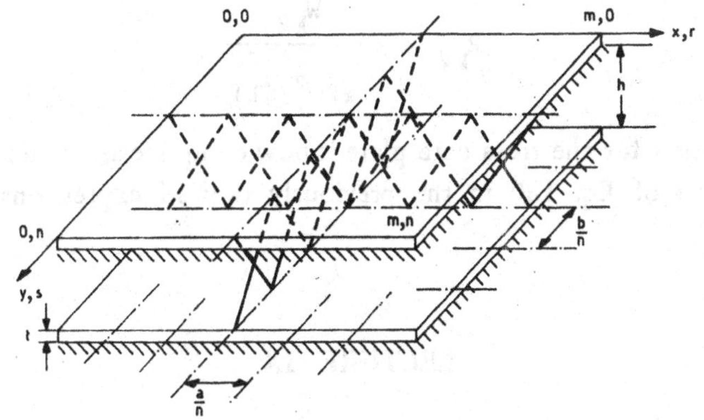

Fig. 8. Discrete Core Sandwich Plate

and the plate, R(r,s) – see Eqs. 123 thru 126 for the plate deflections in terms of the loads on one plate q(x,y) and the out-of-plane interactive forces R(r,s). The interactive force coefficients are determined by matching the plate node deflection to the node deflection of the antisymmetrically loaded two-way truss system with stiff chords which serves as the core. Analysis of such a truss work requires the solution of the same model, Eq. 59d, used to analyze the two dimensional shear grid with the constants dependent upon the properties of the diagonal truss members; i.e.,

$$K = \left(\frac{AE}{\ell_d}\right)\left(\frac{h}{\ell_d}\right)^2 \quad \text{and} \quad \bar{K} = \left(\frac{\overline{AE}}{\bar{\ell}_d}\right)\left(\frac{h}{\bar{\ell}_d}\right)^2$$

The truss solution is

$$w_T(r,s) = \sum_{k=1}^{m-1} \sum_{\ell=1}^{n-1} R_{k\ell} B^T_{k\ell} \sin\frac{k\pi r}{m} \sin\frac{\ell\pi s}{n} \qquad (139a)$$

$$B^T_{k\ell} = \frac{1}{2K\gamma_k + 2\bar{K}\bar{\gamma}_\ell} \qquad (139b)$$

$$\gamma_k = 1 - \cos\frac{k\pi}{m} \qquad \bar{\gamma}_\ell = 1 - \cos\frac{\ell\pi}{n} \qquad (139c,d)$$

Matching coefficients with the double finite series for the plate node deflections – see Eqs. 128 and 129 for the plate on in-line column springs – gives the following results for the interactive force coefficients

$$R_{k\ell} = \frac{W^{o}_{k\ell}}{C_{k\ell} + B^{T}_{k\ell}} \qquad (140)$$

Thus, the solution for the truss core plate is now complete as one needs only to use Eq. 140 in lieu of Eq. 130 in the previously derived expressions for the plate deflection.

LECTURE IX

LATTICED SHELLS

Use of Projected Plane Coordinates. – A single layer latticed shell of two force members is the sort of pure discrete system that appears best suited for a micro field approach. Although such systems are in popular use, their analysis by rational field methods is not well developed; thus, the coverage here is introductory in nature. The object of this section is the statics analysis for the member forces in a latticed shell or curved truss work whose geometry is such that a projection of its nodes on a plane form on evenly spaced pattern of points suitable for description by Cartesian coordinates (but not necessarily orthogonal). That is, the members running in the two coordinate directions and in the diagonal direction from x,y to x + 1, y + 1 (see Fig. 9). Here the discrete coordinates will be labeled x and y to conform with existing publications.(*) (There will be no need for

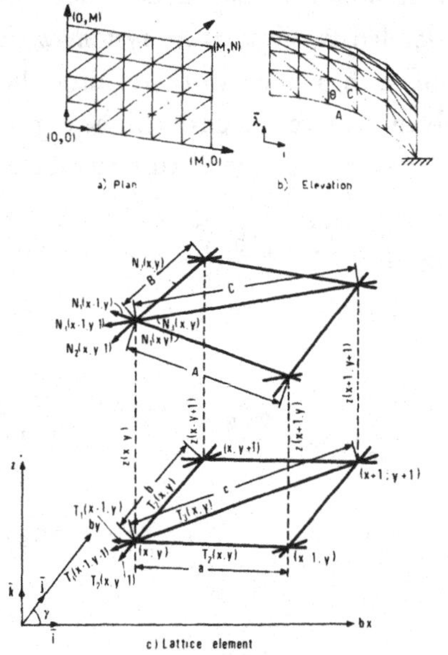

Fig. 9. Latticed Shell and Projection

(*) "On the Statics of Latticed Shells", Donald L. Dean, Publications International Association for Bridge and Structural Engineering, Vol. 25, December 1965, pp. 65-81.

continuous variables in this purely discrete system). The position vector of the nodes is

$$\bar{X} = ax\bar{i} + by\bar{j} + Z(x,y)\bar{k} \tag{141}$$

in which $\bar{k}$ is perpendicular to the plane of $\bar{i}$ and $\bar{j}$. The angle between $\bar{i}$ and $\bar{j}$, ϕ, may be different from $\pi/2$; then the length of the diagonal member is

$$c = (a^2 + b^2 + 2ab\cos\phi)^{\frac{1}{2}} \tag{142a}$$

and the lengths of the surface members are

$$A = [a^2 + (\Delta_x Z)^2]^{\frac{1}{2}} \tag{142b}$$

$$B = [b^2 + (\Delta_y Z)^2]^{\frac{1}{2}} \tag{142c}$$

$$C = [c^2 + (\Delta_u Z)^2]^{\frac{1}{2}} \tag{142d}$$

$$\Delta_u = E_x E_y - 1 \tag{142e}$$

Instead of working with the member forces N_1, N_2 and N_3, the plane components T_1, T_2, and T_3 will be used; i.e.,

$$T_1 = \frac{a}{A} N_1 \;;\quad T_2 = \frac{b}{B} N_2 \;;\quad \text{and}\quad T_3 = \frac{c}{C} N_3 \tag{143}$$

Thus, the general equilibrium equations can be written as :

$$\nabla_x T_1 + \frac{a}{c} \nabla_u T_3 + p_1 = 0 \tag{144a}$$

$$\nabla_y T_2 + \frac{b}{c} \nabla_u T_3 + p_2 = 0 \tag{144b}$$

$$\frac{1}{a} \nabla_x[(\Delta_x Z)T_1] + \frac{1}{b} \nabla_y[(\Delta_y Z)T_2] + \frac{1}{c} \nabla_u[(\Delta_u Z)T_3] + p_z = 0 \tag{144c}$$

in which p_1, p_2 and p_z are components of the load vector

$$\bar{P}(x,y) = p_1\bar{i} + p_2\bar{j} + p_z\bar{k} \tag{145}$$

The out-of-plane equilibrium equation, Eq. 144c, can be rewritten by carrying out the indicated difference of products and substituting Eqs. 144a,b

$$\frac{1}{a}(\Delta_x Z)T_1 + \frac{1}{b}(\Delta_y Z)T_2 + \frac{1}{c}(\Delta_u Z)T_3 - \frac{1}{c}(\nabla_x\nabla_y Z)\nabla_u T_3 = \tag{144d}$$
$$- p_z + \frac{1}{a}(\nabla_x Z)p_1 + \frac{1}{b}(\nabla_y Z)p_2$$

As in the continuum, the three equilibrium equations can be replaced by a single scalar equation through a stress or potential function approach; i.e.,

$$T_1 = \frac{a}{c}\nabla_y\nabla_u F - \nabla_x^{-1} p_1 \tag{145a}$$

$$T_2 = \frac{b}{c}\nabla_x\nabla_u F - \nabla_y^{-1} p_2 \tag{145b}$$

$$T_3 = -\nabla_x\nabla_y F \tag{145c}$$

$$(\Delta_x Z)\nabla_y\nabla_u F - (\Delta_u Z)\nabla_x\nabla_y F + (\Delta_y Z)\nabla_x\nabla_u F + (\nabla_x\nabla_y Z)\nabla_x\nabla_y\nabla_u F = cp_t \tag{145d}$$

$$p_t = - p_z + \frac{1}{a}(\nabla_x Z)p_1 + \frac{1}{b}(\nabla_y Z)p_2 + \frac{1}{a}(\Delta_x Z)\nabla_x^{-1} p_1 + \frac{1}{b}(\Delta_y Z)\nabla_y^{-1} p_2 \tag{145e}$$

As a first example of the use of projected plane coordinates to analyze a latticed shell, consider the case of a hyperbolic paraboloidal lattice with a suitable out-of-plane loading and boundary support forces; i.e., $Z = (f/MN)xy$, $p_1 = p_2 = 0$, $p_z = -p_o$ for $0 < x < M$ and $0 < y < N$, $p_z = -1/2p_o$ on edges except $p_z = 0$ at the unstable joints (M,0) and 0,N. Substitution into Eq. 144d gives

$$(2 - \nabla_u)T_3 = \frac{cMN}{f}p_o \tag{146}$$

The operator in Eq. 146 was discussed in Lecture II (see Eqs. 19 and 21); thus, the solution is

$$T_3(x,y) = \frac{cMN}{2f}p_o[1 + (-1)^y G(x-y)] \tag{147a}$$

Substitution into Eqs. 144a and 144b give

$$\nabla_x T_1 = -\frac{aMN}{f}p_o(-1)^y G(x-y) \tag{147b}$$

$$\nabla_y T_2 = -\frac{bMN}{f} p_o (-1)^y G(x-y) \tag{147c}$$

The homogeneous part of the solution for T_3 alternates in sign and maintains a constant magnitude as one progresses from one member to the next along a given diagonal. This, plus the known solution for the continuous case, suggests that suitable boundary conditions may be met with $G(x-y) = 0$. This trial solution will be tested below.

$$T_3 = \frac{cMN}{2f} p_o \quad , \quad T_1 = G_1(y) \quad , \quad T_2 = G_2(x) \tag{148a,b,c}$$

to satisfy 1) zero net $\bar{j}$ force along $y = 0$ and 2) zero net $\bar{i}$ force along $x = 0$,

$$T_2(x,0) + \frac{b}{c} T_3(x,0) = 0 \qquad T_1(0,y) + \frac{a}{c} T_3(0,y) = 0 \tag{149a,b}$$

$$T_1 = -\frac{aMN}{2f} p_o \quad \text{and} \quad T_2 = -\frac{bMN}{2f} p_o \tag{150a,b}$$

The reader should study the significance of this solution from the standpoint of the type of reactions required along the edges and the corners.

As a second example of the use of projected plane coordinates, consider the case of an elliptic paraboloidal lattice loaded normal to the reference plane; i.e.,

$$Z = \frac{d}{M^2} x^2 + \frac{e}{N^2} y^2 \quad , \quad p_1 = p_2 = 0 \tag{151a,b,c}$$

Substitution into Eq. 145d gives:

$$\left(\frac{e}{N^2} \Delta_x + \frac{d}{M^2} \Delta_y\right) F(x,y) = -\frac{c}{2} p_z (x+1, y+1) \tag{152}$$

This is nearly the same model as for the shear grid, Eq. 59d, thus a double series solution can be easily written provided the boundary conditions can be satisfied by the terms of the series. As in the previous problem, we will study the case of zero normal edge forces; i.e., Eqs. 149a,b plus the following:

$$T_2(x,N) + \frac{b}{c} T_3(x,N) = 0 \quad \text{and} \quad T_1(M,y) + \frac{a}{c} T_3(M,y) = 0 \tag{153a,b}$$

Substitution of the stress function definition, Eq. 145a,b and c gives :

$$\text{(154a,b)} \qquad \Delta_y F\left(\frac{-1}{M-1}, y-1\right) = \Delta_x F\left(x-1, \frac{-1}{N-1}\right) = 0$$

or

$$\text{(154c,d)} \qquad F\left(\frac{-1}{M-1}, y-1\right) = F\left(x-1, \frac{-1}{N-1}\right) = 0$$

Thus, the solution for F(x,y) can be written in the form

$$\text{(155a)} \qquad F(x,y) = \sum_{k=1}^{M-1} \sum_{\ell=1}^{N-1} A_{k\ell} \sin \frac{k\pi}{M}(x+1) \sin \frac{\ell\pi}{N}(y+1)$$

$$\text{(155b)} \qquad p_z(x,y) = \sum_{k=1}^{M-1} \sum_{\ell=1}^{N-1} P_{k\ell} \sin \frac{k\pi x}{M} \sin \frac{\ell\pi y}{N}$$

$$\text{(155c)} \qquad A_{k\ell} = \frac{\frac{1}{4} c P_{k\ell}}{\frac{e}{N^2}\gamma_k + \frac{d}{M^2}\bar{\gamma}_\ell}$$

$$\text{(155d,e)} \qquad \gamma_k = 1 - \cos \frac{k\pi}{M} \quad ; \quad \bar{\gamma}_\ell = 1 - \cos \frac{\ell\pi}{N}$$

This completes the solution for the elliptic paraboloidal lattice as the various stress components can be obtained by operating on Eqs. 155 in accordance with Eqs. 145 and 143.

Concepts of Difference Geometry. – The geometry of many useful surfaces is such that it is not feasible to construct a lattice whose nodes lie on the surface and also satisfy the conditions required for analysis by a projected plane approach. For the more general shapes, some form of discrete surface coordinate system is required analogous to the continuous surface coordinates used to analyze whole shells. This requires that one introduce and develop some basic definitions and concepts of difference geometry which is as fundamental to latticed shell theory as differential geometry is to continuous shell theory.

It is assumed that the position vector for the nodes of a latticed surface is known as a function of two parameters α_1 and α_2 which form the surface coordinates; i.e., α_1 = constant defines the α_2 polygons of members on the surface and α_2 = constant defines the α_1 surface polygons, see Fig. 10. (Note that the

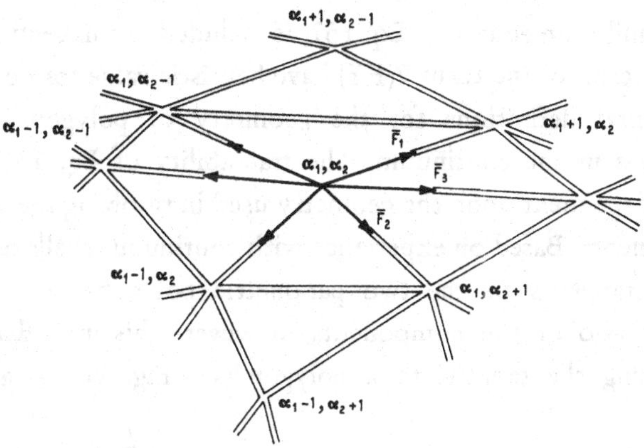

Fig. 10. Lattice Shell Element

coordinate curves of the continuum become polygons for a latticed surface).

For a triply laced surface, there is a third set of dependent polygons selected here so that their chords connect α_1, α_2 with $\alpha_1 + 1, \alpha_2 + 1$. The three unit chord or member vectors are

$$\bar{r}_1 = \frac{1}{L_1} \Delta_1 \bar{X}(\alpha_1, \alpha_2) \; ; \; \bar{r}_2 = \frac{1}{L_2} \Delta_2 \bar{X}(\alpha_1, \alpha_2) \; ; \; \bar{r}_3 = \frac{1}{L_3} \Delta_3 \bar{X}(\alpha_1, \alpha_2) \qquad \text{(156a,b,c)}$$

and the joint equilibrium, in terms of the member forces F_1, F_2 and F_3, is expressed by the following vector difference equation:

$$\nabla_1 (F_1 \bar{r}_1) + \nabla_2 (F_2 \bar{r}_2) + \nabla_3 (F_3 \bar{r}_3) + \bar{P} = 0 \qquad (157)$$

in which the forward and backward differences Δ_1, Δ_2, ∇_1 and ∇_2 are as normally defined and the dependent operators are defined as:

$$\nabla_3 = 1 - E_1^{-1} E_2^{-1} = \mathbf{M}_1 \nabla_2 + \mathbf{M}_2 \nabla_1 \qquad \text{(158a,b)}$$

$$\Delta_3 = E_1 E_2 - 1 = N_1 \Delta_2 + \Delta_2 N_2 \qquad \text{(158c,d)}$$

The equilibrium equation, Eq. 157, is included in this section on geometry because the three terms of the form $\nabla(F\,\bar{r})$ have heuristic value for determining some of the fundamental definitions for the geometry of polygons – analogous to geometry of curves in the continuum. The tractability of Eq. 157 for a specific surface is highly dependent upon the geometry used in resolving the vector equation into scalar components. Based on experience with continuous shells one would think in terms of the tangents to the two parametric polygons, α_1 = constant and α_2 = constant as two of the components; however, this immediately raises the question of defining the tangent to a polygon (see Fig. 11). A great variety of

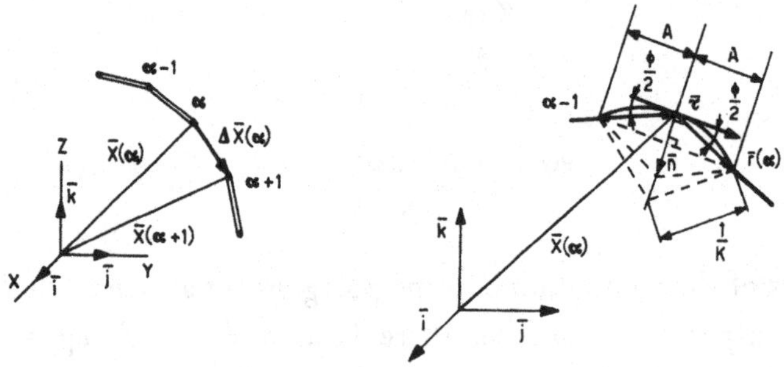

Fig. 11 a, b. Polygon Geometry

possible definitions degenerate to the tangent of a curve as the distance between nodes approaches zero but only one definition has stood the tests of use, symmetry and elegance(*) and it is the one that arises naturally as one studies the difference of the product of $F\bar{r}$ shown in Eq. 157.

$$\text{(159a)} \qquad \nabla(F(\alpha)\bar{r}(\alpha)) = [\nabla F(\alpha)][H\bar{r}(\alpha)] + [HF(\alpha)][\nabla\bar{r}(\alpha)]$$

in which it is noted that $[H\bar{r}(\alpha)]\,.\,[\nabla\bar{r}(\alpha)] = 0$. This led to the adoption of the directions of $H\bar{r}(\alpha)$ and $\nabla\bar{r}(\alpha)$ respectively as the directions of polygon tangent, $\bar{t}$, and normal, $\bar{n}$; i.e.,

$$\text{(159b)} \qquad \nabla[F(\alpha)\bar{r}(\alpha)] = [\nabla F(\alpha)][A\bar{t}] + [HF(\alpha)][K\bar{n}]$$

(*) Note that $\bar{t}$ is parallel to the carom wall thru α for points $\alpha+1$ and $\alpha-1$; i.e. $\bar{t}$ bisects angle between $\bar{r}(\alpha)$ and $\bar{r}(\alpha-1)$.

in which the length factor A and the polygon curvature K are the magnitudes of $\mathcal{H}\bar{r}$ and $\nabla\bar{r}$ respectively:

$$A = \left| \mathcal{H}\bar{r}(\alpha) \right| = \cos\frac{\phi}{2} \tag{159c}$$

$$K = \left| \nabla\bar{r}(\alpha) \right| = 2\sin\frac{\phi}{2} \tag{159d}$$

and ϕ is the deflection angle between successive chords:

$$\cos\phi = [\bar{r}(\alpha)]\cdot[\bar{r}(\alpha-1)] \tag{159e}$$

If required, the polygon binormal, $\bar{b}$, is defined as $\bar{b} = \bar{t} \times \bar{n}$.

The polygon length factor A plays the same role as the length factor for the differential arc length of a curve in the continuum and extends in the same manner to the coefficients of the first fundamental form for a latticed surface; i.e.,

$$(A_{ij})^2 = (\mathcal{H}_i\bar{r}_i)\cdot(\mathcal{H}_j\bar{r}_j) \qquad i,j = 1,2 \tag{160a}$$

which relates to the angle θ between parametric polygons as follows:

$$\cos\theta = \frac{(A_{12})^2}{A_{11}A_{22}} \tag{160b}$$

$$A_s^2 = (A_{11})^2(A_{22})^2 - (A_{12})^4 \tag{160c}$$

The surface normal is defined as

$$\bar{N} = \frac{\bar{t}_1 \times \bar{t}_2}{\sin\theta} = \frac{(\mathcal{H}_1\bar{r}_1)\times(\mathcal{H}_2\bar{r}_2)}{A_s} \tag{160d,e}$$

Note that $\bar{N}$ is not, in general, parallel to either of the normals to the parametric surface polygons, $\bar{n}_1$ or $\bar{n}_2$. Work to establish the coefficients of the second fundamental form for latticed surfaces is not yet complete. From the above, it is seen that the force exerted on a node by the two two-force members of a single polygon has the following components:

1) $A\nabla F$ – tangential component parallel to the polygon tangent;
2) $(K\mathcal{H}F)\cos\gamma$ – normal curvature component, parallel to the surface normal, in which $\cos\gamma = \bar{n}\cdot\bar{N}$; and

3) (K И F) sin γ – geodesic curvature component, perpendicular to polygon tangent in the surface tangent plane. (Geodesic polygons are those with zero geodesic curvature or $\bar{n}_{\iota} = \bar{N}$ and $\gamma = 0$). These concepts permit the writing of general equilibrium equations, but it is usually just as convenient to apply them directly to the specific shell in question as it is to modify the general equations.

Application of Discrete Surface Coordinates. – The object of this section is to apply the discrete geometry concepts on the previous section to the stress an deflection analysis of the single layered cylindrical lattice shown in Fig. 12.

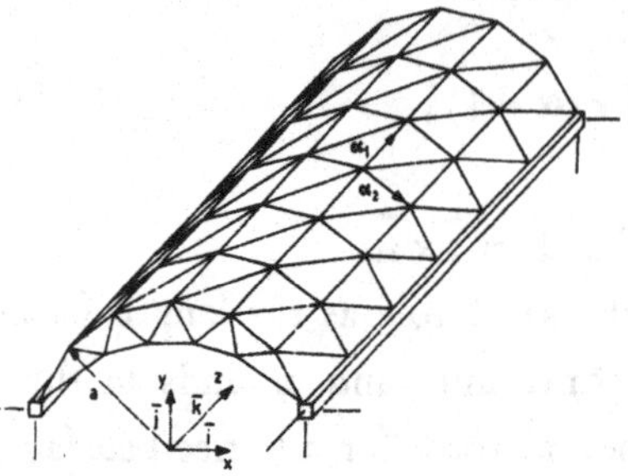

Fig. 12. Latticed Cylindrical Shell

The latticed shell to be considered in detail is the cylindrical truss work shown in Fig. 12. The α_1 "polygons" (α_2 = constant) are the straight longitudinal members. The α_2 coordinates (α_1 = constant) are along the circular cross-sectional polygons. The diagonal members ($\alpha_1 - \alpha_2$ = constant) are the helical polygons, which are geodesic. The position vectors and other required geometrical quantities are:

$$\bar{X}(\alpha_1, \alpha_2) = a(\sin\phi\alpha_2)\bar{i} + a(\cos\phi\alpha_2)\bar{j} + L_1\alpha_1\bar{k}$$

$$L_2 = a\,K_2 \ ; \quad K_2 = 2\sin\frac{\phi}{2} \ ; \quad L_3 = [L_1^2 + L_2^2]^{\frac{1}{2}}$$

(161)

$$\bar{r}_1 = \bar{t}_1 = \bar{k} \ ; \quad \bar{r}_2 = \cos\phi\left(\alpha_2 + \frac{1}{2}\right)\bar{i} - \sin\phi\left(\alpha_2 + \frac{1}{2}\right)\bar{j}$$

$$\bar{t}_2 = (\cos\phi\alpha_2)\bar{i} - (\sin\phi\alpha_2)\bar{j}$$

$$A_{11} = A_1 = 1.0 \ ; \quad A_{22} = A_2 = \cos\frac{\phi}{2} \ ; \quad A_{12} = 0$$

$$\bar{n}_2 = \bar{N} = -(\sin\phi\alpha_2)\bar{i} - (\cos\phi\alpha_2)\bar{j}$$

$$\bar{r}_3 = \frac{1}{L_3}(L_2\bar{r}_2 + L_1\bar{k})$$

$$\bar{t}_3 = \frac{1}{A_3 L_3}[A_2 L_2 \bar{t}_2 + L_1 \bar{k}]$$

$$A_3 = \frac{1}{L_3}[A_2^2 L_2^2 + L_1^2]^{\frac{1}{2}}$$

$$K_3 = \frac{L_2}{L_3} K_2 \; ; \; \bar{n}_3 = \bar{n}_2 = \bar{N}$$

Use of these relations and the basic concepts developed in the previous section yield the following scalar difference equations for the $\bar{t}_1$, $\bar{t}_2$, $\bar{N}$ force components.

$$\begin{bmatrix} \nabla_1 & 0 & \frac{L_1}{L_3}\nabla_3 \\ 0 & \nabla_2 & \frac{L_2}{L_3}\nabla_3 \\ 0 & \mathcal{H}_2 & \frac{L_2}{L_3}\mathcal{H}_3 \end{bmatrix} \begin{bmatrix} F_1 \\ F_2 \\ F_3 \end{bmatrix} = - \begin{bmatrix} P_1 \\ \frac{1}{A_2}P_2 \\ \frac{1}{K_2}P_n \end{bmatrix} \qquad (162)$$

Assuming sufficient support conditions along the latticed shell boundaries to satisfy basic stability conditions, Eqs. 162 can be solved directly for the member forces, F_1, F_2 and F_3 in terms of the given joint loads P_1, P_2 and P_n and this will be done subsequently for the special case of a uniform vertical "dead" load. Before proceeding with such a "membrane" analysis, however, it is helpful to extend the mathematical model to one that can be used to find joint displacements by substituting the force-displacement relations into the equilibrium equations. The force-displacement relations are as follows:

$$F_1 = Q_1 \Delta_1 V_1 \qquad (163a)$$

$$F_2 = Q_2 [A_2 \Delta_2 V_2 - K_2 N_2 V_n] \qquad (163b)$$

$$F_3 = \frac{Q_3}{L_3}[L_1 \Delta_3 V_1 + L_2 A_2 \Delta_3 V_2 - L_2 K_2 N_3 V_n] \qquad (163c)$$

In which V_1, V_2 and V_n are the joint displacements in the $\bar{t}_1$, $\bar{t}_2$ and $\bar{N}$ directions respectively and Q_1, Q_2, and Q_3 are the "spring" constants, product of cross-sectional area and elastic modulus divided by length, for the truss members parallel to $\bar{r}_1$, $\bar{r}_2$ and $\bar{r}_3$ respectively. Substitution of these relations into the equilibrium equations yields the following model for the joint displacements:

$$(164)\qquad \begin{bmatrix} C_{11} & A_2 C_{21} & -K_2 C_{31} \\ C_{21} & A_2 C_{22} & -K_2 C_{32} \\ C_{31} & A_2 C_{32} & -K_2 C_{33} \end{bmatrix} \begin{bmatrix} V_1 \\ V_2 \\ V_3 \end{bmatrix} = \frac{-1}{\bar{Q}_3} \begin{bmatrix} P_1 \\ \frac{1}{A_2} P_2 \\ \frac{1}{K_2} P_n \end{bmatrix}$$

in which the elements of the operator matrix are given by

$$(165a)\qquad C_{11} = \frac{Q_1}{\bar{Q}_3} \Delta_1 + \frac{L_1}{L_2} \Delta_3$$

$$(165b,c)\qquad C_{21} = \Delta_3 \; ; \quad C_{31} = \boxminus_3$$

$$(165d)\qquad C_{22} = \frac{Q_2}{\bar{Q}_3} \Delta_2 + \frac{L_2}{L_1} \Delta_3$$

$$(165e)\qquad C_{32} = \frac{Q_2}{\bar{Q}_3} \boxminus_2 + \frac{L_2}{L_1} \boxminus_3$$

$$(165f)\qquad C_{33} = \frac{1}{4} \frac{Q_2}{\bar{Q}_3} (\Delta_2 + 4) + \frac{1}{4} \frac{L_2}{L_1} (\Delta_3 + 4)$$

The operator determinant is found to be

$$(166)\qquad | C_{ij} | = - K_2 A_2 \left(\frac{L_2}{L_1} \right) \left(\frac{Q_1 Q_2}{\bar{Q}_3^2} \right) \Delta_1^2$$

From Eq. 166 it is seen that the total solution will have two arbitrary functions each at $\alpha_1 = 0$ and $\alpha_1 = m$ but we will be unable to satisfy arbitrary conditions at the side boundaries.

The procedure here will be to solve the statics model Eq. 162 for F_1, F_2 and F_3 and then, as a second step substitute into Eq. 163 to solve for V_1, V_2 and V_n.

The operator determinant of the 2nd and 3rd rows of Eq. 162 is $-(L_2/L_3) E_2^{-1} \nabla_1$; thus, Cramer's rule allows one to uncouple the model as follows:

$$\nabla_1 F_3 = \frac{L_3}{L_2}\left[\frac{1}{K_2}\Delta_2 P_n - \frac{1}{A_2} N_2 P_2\right] \tag{167a}$$

$$F_2 = -\left[\frac{L_2}{L_3} F_3 + \frac{1}{2A_2} P_2 + \frac{1}{K_2} P_n\right] \tag{167b}$$

$$\nabla_1 F_1 = -\frac{L_1}{L_3}\nabla_3 F_3 - P_1 \tag{167c}$$

For the special case of a uniform gravity load, i.e., $\bar{P} = -P_o\bar{j}$ (or $-1/2P_o\bar{j}$ on boundary nodes) the loading terms are: $P_1 = 0$, $P_2 = P_o \sin\phi\,\alpha_2$ and $P_n = P_o\cos\phi\,\alpha_2$.

Substitution into Eq. 162 and carrying out the indicated inverse operations yields the following general statics solution:

$$F_3(\alpha_1,\alpha_2) = -\frac{2L_3 P_o}{L_2}\alpha_1 \sin\phi\left(\alpha_2 + \frac{1}{2}\right) + G_1(\alpha_2) \tag{168a}$$

$$F_2(\alpha_1,\alpha_2) = 2P_o\alpha_1 \sin\phi\left(\alpha_2 + \frac{1}{2}\right) - \frac{P_o}{\sin\phi}\cos\phi\left(\alpha_2 - \frac{1}{2}\right) - \frac{L_2}{L_3}G_1(\alpha_2) \tag{168b}$$

$$F_1(\alpha_1,\alpha_2) = 2\frac{P_o L_1}{L_2}\alpha_1\left[\alpha_1 \sin\frac{\phi}{2}\cos\phi\ \alpha_2 + \cos\frac{\phi}{2}\sin\phi\,\alpha_2\right] - \frac{L_1}{L_3}\alpha_1\nabla_2 G_1(\alpha_2) + G_2(\alpha_2) \tag{168c}$$

The arbitrary functions $G_1(\alpha_2)$ and $G_2(\alpha_2)$ are determined so as to satisfy the conditions of thin end arches; i.e.,

$$F_1(0,\alpha_2) + \frac{L_1}{L_3}F_3(0,\alpha_2) = 0 \tag{169a}$$

$$F_1(m,\alpha_2) + \frac{L_1}{L_3}F_3(m,\alpha_2) = 0 \tag{169b}$$

(170a) $$G_2(\alpha_2) = -\frac{L_1}{L_3} G_1(\alpha_2)$$

(170b) $$G_1(\alpha_2) = \frac{P_o L_3}{L_2} (m-1) \sin\phi \left(\alpha_2 + \frac{1}{2}\right)$$

Thus, the complete statics solution can be written:

$$F_3(\alpha_1, \alpha_2) = \frac{P_o L_3}{L_2} (m - 1 - 2\alpha_1) \sin\phi \left(\alpha_2 + \frac{1}{2}\right)$$
(171a)

$$F_2(\alpha_1, \alpha_2) = -P_o (m - 1 - 2\alpha_1) \sin\phi \left(\alpha_2 + \frac{1}{2}\right) - P_o \frac{\cos\phi \left(\alpha_2 - \frac{1}{2}\right)}{\sin\phi}$$
(171b)

$$F_1(\alpha_1, \alpha_2) = -\frac{P_o L_1}{L_2} \left[(m - 1 - 2\alpha_1) \sin\phi \left(\alpha_2 + \frac{1}{2}\right) + 2\alpha_1 (m - \alpha_1) \sin\frac{\phi}{2} \cos\phi\, \alpha_2 \right]$$
(171c)

Note that the solution satisfies the following approximation to the condition of a thin edge beam:

(172) $$F_2(\alpha_1, n-1) + \frac{L_2}{L_3} F_3(\alpha_1 - 1, n-1) = \frac{P_o}{2} \left(\frac{\sin\phi n}{\cos\frac{\phi}{2}} - \frac{\cos\phi n}{\sin\frac{\phi}{2}} \right)$$

For exact satisfaction of the condition of zero external $\bar{t}_2$ component along $\alpha_2 = n$, the second term on the right should be zero, which is the case for a semi-circular cross-section ($\phi_n = \pi/2$). However, the general statics solution does not permit specification of conditions along the side boundary, so some tangential side support must be furnished for sectional shells.

Substitution of the solution for member forces into the force displacement relations, Eq. 163, gives the model for the second stage of the solution, that of finding node displacements. The first relation gives V_1 directly. The second and third relations can be solved simultaneously for V_2 and V_3 in a manner closely analogous to that used to find F_2 and F_3 from the 2nd and 3rd rows of Eq. 162. Here the operator determinant is $K_2 A_2 E_2 \Delta_1$. Thus, the general solution for the node displacements is :

$$V_1(\alpha_1,\alpha_2) = \Delta_1^{-1}\left(\frac{F_1}{Q_1}\right) + G_3(\alpha_2) \tag{173a}$$

$$V_1 = \frac{P_o L_1}{Q_1 L_2}\alpha_1\left[(m-\alpha_1)\sin\phi\left(\alpha_1+\frac{1}{2}\right)+\frac{1}{3}(\alpha_1-1)(3m+1-2\alpha_1)\sin\frac{\phi}{2}\cos\phi\,\alpha_2\right]$$

$$+\,G_3(\alpha_2) \tag{173b}$$

$$A_2 V_2(\alpha_1,\alpha_2) = \Delta_1^{-1}\nabla_2\left(\frac{L_3F_3}{L_2Q_3}-\frac{F_2}{Q_2}-\frac{L_1}{L_2}\Delta_3 V_1\right)-\frac{F_2}{2Q_2}+N_2G_4(\alpha_2) \tag{173c}$$

$$K_2 V_n(\alpha_1,\alpha_2) = \Delta_1^{-1}\nabla_2\left(\frac{L_3F_3}{L_2Q_3}-\frac{F_2}{Q_2}-\frac{L_1}{L_2}\Delta_3 V_1\right)-\frac{F_2}{Q_2}+\Delta_2G_4(\alpha_2) \tag{173d}$$

Carrying out the indicated operations in Eqs. 173c and 173d, leads to the following general expressions for V_2 and V_n .

$$V_2 = \left[\left(\frac{P_oL_3}{Q_3L_2^2}+\frac{P_o}{Q_2}+\frac{P_oL_1^2}{Q_1L_2^2}\cos\phi\right)\alpha_1(m-\alpha_1)+\frac{P_o}{2Q_2}\left(m-\frac{\cos\phi}{1+\cos\phi}\right)\right.$$

$$\left.-\frac{1-\cos\phi}{6}\frac{P_oL_1^2}{Q_1L_2^2}\alpha_1(\alpha_1^2-1)(2m-\alpha_1)\right]\sin\phi\,\alpha_2$$

$$+\left[\frac{P_o}{2Q_2}\left(m+\frac{(4\alpha_1+1)\cos\phi-2\alpha_1}{1-\cos\phi}\right)+\frac{1+\cos\phi}{3}\frac{P_oL_1^2}{Q_1L_2^2}\left(\alpha_1^2(3m-2\alpha_1)-\alpha_1\right)\right](*)$$

$$(*) = \tan\frac{\phi}{2}\cos\phi\,\alpha_2$$

$$-\frac{L_1}{L_2}\alpha_1\nabla_2G_3(\alpha_2)+N_2G_4(\alpha_2) \tag{173e}$$

$$V_n = \left[\left(\frac{P_oL_3^2}{Q_3L_2^2}+\frac{P_o}{Q_2}+\frac{P_oL_1^2}{Q_1L_2^2}\cos\phi\right)\alpha_1(m-\alpha_1)+\frac{P_o}{2Q_2}\left(m+\frac{\cos\phi}{1-\cos\phi}\right)\right.$$

$$\left.-\frac{1-\cos\phi}{6}\frac{P_oL_1^2}{Q_1L_2^2}\alpha_1(\alpha_1^2-1)(2m-\alpha_1)\right]\cos\phi\,\alpha_2$$

$$+\left[\frac{P_o}{2Q_2}\left(m-\frac{(4\alpha_1+1)\cos\phi+2\alpha_1}{1+\cos\phi}\right)-\frac{1-\cos\phi}{3}\frac{P_oL_1^2}{Q_1L_2^2}\left(\alpha_1^2(3m-2\alpha_1)-\alpha_1\right)\right](*)$$

$$(*) = \cot\frac{\phi}{2}\sin\phi\,\alpha_2$$

$$-\frac{L_1}{L_2}\alpha_1\nabla_2G_3(\alpha_2)+\Delta_2G_4(\alpha_2) \tag{173f}$$

The arbitrary boundary functions $G_3(\alpha_2)$ and $G_4(\alpha_2)$ are determined from the additional condition imposed by end arches with high inplane stiffnesses.

$$V_2(0,\alpha_2) = V_2(m,\alpha_2) = 0 \tag{174a,b}$$

$$G_3(\alpha_2) = \left(\frac{P_o L_1}{Q_1 L_2}\frac{m^2-1}{3} - \frac{P_o L_2}{Q_2 L_1}\frac{1-2\cos\phi}{\sin^2\phi}\right)\cos\frac{\phi}{2}\sin\phi\alpha_2 + \frac{P_o L_1}{Q_1 L_2}\frac{m(m^2-1)}{6}\sin\frac{\phi}{2}\cos\phi\alpha_2 \tag{175a}$$

$$G_4(\alpha_2) = -\frac{P_o}{2A_2 Q_2}(m\sin\phi\alpha_2 + \cot\phi\cos\phi\alpha_2) \tag{175b}$$

This completes the stress and deflection analysis of the cylindrical lattice shown in Fig. 12.

LECTURE X

THREE–DIMENSIONAL SYSTEMS AND SPECIAL PROBLEMS

Three-Dimensional Networks. – Although continuum mechanics is much better developed than discrete field mechanics, there is available only a small number of useful three dimensional solutions in the continua; however, it is relatively easy to write discrete field solutions to a variety of functional three-dimensional lattices and two such systems will be analyzed briefly in this section. A three dimensional system is one for which three coordinates are required to locate a typical node. This is as distinct from a latticed shell for example which is a space structure of only two dimensions as regards the number of surface or projected plane coordinates required to define a typical node. (Latticed shells are often improperly designated as three dimensional structures).

As the first example of a three dimensional system, consider a regular three dimensional network of stretched cables fixed on all six faces of the circumscribing rectangular solid and loaded parallel to one set of cables. The three coordinates have the range $r_i = 0,(1),N_i$, $i = 1,2,3$. The cable tensions are T_i and the uniform cable spacings in the coordinate directions are a_i. The loading is $P_3(r_1,r_2,r_3)$ and the desired solution is for the corresponding node deflections $w_3(r_1,r_2,r_3)$ both of which can be expressed as triple sine series for the case of zero boundary displacements; i.e.,

$$\begin{bmatrix} P_3(r_1,r_2,r_3) \\ \\ w_3(r_1,r_2,r_3) \end{bmatrix} = \sum_{k_1}\sum_{k_2}\sum_{k_3} \begin{bmatrix} C(k_1,k_2,k_3) \\ \\ A(k_1,k_2,k_3) \end{bmatrix} \sin\frac{k_1\pi r_1}{N_1}\sin\frac{k_2\pi r_2}{N_2}\sin\frac{k_3\pi r_3}{N_3} \tag{176a}$$

$$k_i = 1,(1),N_i - 1 \tag{176b}$$

The mathematical model is written by summing the resistance of the nets in the r_1,r_2 planes to out-of-plane displacement and the resistance of the r_3 cables to differential stretching; i.e.

$$\left[\frac{T_1}{a_1}\Delta\!\!\!\!\nabla_1 + \frac{T_2}{a_2}\Delta\!\!\!\!\nabla_2 + \frac{(AE)_3}{a_3}\Delta\!\!\!\!\nabla_3\right] w_3(r_1,r_2,r_3) = -P_3(r_1,r_2,r_3) \tag{177}$$

in which $\Delta\!\!\!\!\nabla_i$ denote second central differences with respect to the corresponding coordinates r_i. The solution for the displacement coefficients $A(k_1,k_2,k_3)$ in terms of the load coefficients $C(k_1,k_2,k_3)$ is found by substituting Eq. 176 into Eq. 177 and matching coefficients; i.e.,

$$A(k_1,k_2,k_3) = \frac{\frac{1}{2}C(k_1,k_2,k_3)}{\frac{T_1}{a_1}\gamma_1(k_1) + \frac{T_2}{a_2}\gamma_2(k_2) + \frac{(AE)_3}{a_3}\gamma_3(k_3)} \tag{178a}$$

in which

$$\gamma_i(k_i) = 1 - \cos\frac{k_i\pi}{N_i} \tag{178b}$$

For a second and closely similar three dimensional model, consider the system comprised of a stack of shear grids of the type shown in Fig. 13 separated by springs of constant σ . Using coordinates as for the network problem, the model is

$$(K\Delta\!\!\!\!\nabla_1 + \bar{K}\Delta\!\!\!\!\nabla_2 + \sigma\Delta\!\!\!\!\nabla_3)w(r_1,r_2,r_3) = -P(r_1,r_2,r_3) \tag{179a}$$

$$\nabla_1 w(0,r_2,r_3) = \Delta_1 w(N_1,r_2,r_3) = \nabla_2 w(r_1,0,r_3) = \Delta_2 w(r_1,N_2,r_3) = 0 \tag{179b}$$

$$\nabla_3 w(r_1,r_2,0) = w(r_1,r_2,N_3) = 0 \tag{179c}$$

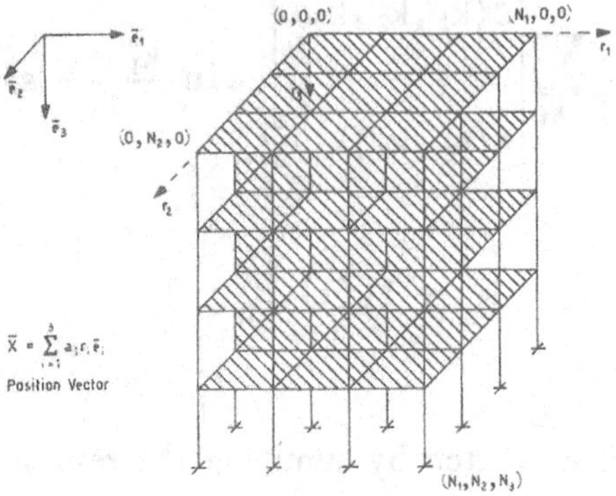

Fig. 13. A Three Dimensional Frame

For this case, the solution and load series have the form

$$\begin{bmatrix} P(r_1, r_2, r_3) \\ \\ w(r_1, r_2, r_3) \end{bmatrix} = \sum_{k_1}\sum_{k_2}\sum_{k_3} \begin{bmatrix} C(k_1, k_2, k_3) \\ \\ A(k_1, k_2, k_3) \end{bmatrix} \cos \lambda_k^1\left(r_1 + \frac{1}{2}\right) \cos \lambda_k^2\left(r_2 + \frac{1}{2}\right) (*)$$

(180a)

$$(*) = \cos \lambda_k^3\left(r_3 + \frac{1}{2}\right)$$

(180b) $k_1 = 0, (1), N_1 \;;\quad k_2 = 0, (1), N_2 \;;\quad k_3 = 0, (1), N_3 - 1$

(180c) $$\lambda_k^1 = \frac{k_1 \pi}{N_1 + 1} \;;\quad \lambda_k^2 = \frac{k_2 \pi}{N_2 + 1} \;;\quad \lambda_k^3 = \frac{k_3 + \frac{1}{2}}{N_3 + \frac{1}{2}} \pi$$

The solution is found as in the previous problem; i.e.,

(181a) $$A(k_1, k_2, k_3) = \frac{\frac{1}{2} C(k_1, k_2, k_3)}{K\gamma_1(k_1) + \bar{K}\gamma_2(k_2) + \sigma\gamma_3(k_3)}$$

(181b) $$\gamma_i(k_i) = 1 - \cos \lambda_k^i$$

Three-Dimensional Frameworks. – For a second class of three-dimensional structures, consider the regular flexural framework showh in Fig. 13. The frame is proportioned so that the effects of member shortening are negligible and loaded in the r_3 direction only and in a symmetric fashion such that none of the vertical frames tend to displace(*).

If only "St. Venant" torsion is considered (no warping and no coupling between bending and torsion in a member element), then the moment-rotation relation can be written compactly as follows:

$$M_{ij}(r_1, r_2, r_3) = (bK)_{ij} \left[(\Delta_i + \gamma_{ij}) \theta_j (r_1, r_2, r_3) \right] \tag{182a}$$

$$E_i^{-1} M'_{ij} (r_i, r_2, r_3) = (bK)_{ij} \left[(\gamma_{ij} - \nabla_i) \theta_j (r_1, r_2, r_3) \right] \tag{182b}$$

in which M_{ij} are the moments at the reference end of a member parallel to the $\bar{e}_i$ axis with a rotation direction clockwise about the $\bar{e}_j$ axis; M'_{ij} is the corresponding moment at the opposite end of the same member element; $(bK)_{ij}$ is a bending stiffness coefficient of the member parallel to the $\bar{e}_i$ axis in the $\bar{e}_j$ direction, i.e., for prismatic members

$$(bK)_{ij} = \frac{2(EI)_{ij}}{a_i} \quad \text{for } i \neq j, \text{ and } (bK)_{ii} = \frac{(GJ)_i}{a_i} \quad \text{for } i = j ;$$

γ_{ij} is as defined for Eq. 43 for $i \neq j$ (equal to 3 for prismatic members) and for $i = j$, $\gamma_{ii} = 0$.

The three uncoupled equilibrium equations are:

$$\sum_{i=1}^{3} (bK)_{ij} (\Delta_i + 2\gamma_{ij}) \theta_j (r_1, r_2, r_3) = M_j^o (r_1, r_2, r_3) \qquad j = 1,2,3 \tag{183}$$

in which M_j^o are the effective applied joint moments about the $\bar{e}_j$ axes as defined for Eq. 44 (here $M_3^o = 0$ due to absence of horizontal loads; thus, $\theta_3 = 0$).

For the boundary condition of one-half frame stiffnesses on the sides and top and fixed bases at the bottom, the solution and load terms can be expanded as follows:

(*) Such restrictions are a convenience for this presentation not a necessity. Solutions are available for regular flexural frameworks which account for joint displacements as well as rotations(**).

(**) "Discrete Field Analysis of Regular Three Dimensional Structural Frames", Anooshiravan Askari, Civil Engineering Doctor of Philosophy Thesis, North Carolina State University, Raleigh, North Carolina, 1969.

$$\begin{bmatrix} M_1^o(r_1,r_2,r_3) \\ \\ \theta_1(r_1,r_2,r_3) \end{bmatrix} = \sum_{k_1}\sum_{k_2}\sum_{k_3} \begin{bmatrix} \bar{\underline{M}}_1^o(k_1,k_2,k_3) \\ \\ \bar{\underline{\theta}}_1(k_1,k_2,k_3) \end{bmatrix} \cos\lambda_k^1 r_1 \cos\lambda_k^2 r_2 \cos\lambda_k^3 r_3$$

(184a)

in which

(184b) $k_1 = 0,(1),N_1 \;;\; k_2 = 1,(2),N_2 \;;\; k_3 = 0,(1),N_3 - 1$

(184c) $\lambda_k^1 = \frac{k_1\pi}{N_1} \;,\; \lambda_k^2 = \frac{k_2\pi}{N_2} \;,\; \lambda_k^3 = \frac{k_3 + \frac{1}{2}}{N_3}\pi$

$$\begin{bmatrix} M_2^o(r_1,r_2,r_3) \\ \\ \theta_1(r_1,r_2,r_3) \end{bmatrix} = \sum_{k_1}\sum_{k_2}\sum_{k_3} \begin{bmatrix} \bar{\underline{M}}_2^o(k_1,k_2,k_3) \\ \\ \bar{\underline{\theta}}_2(k_1,k_2,k_3) \end{bmatrix} \cos\lambda_k^1 r_1 \cos\lambda_k^2 r_2 \cos\lambda_k^3 r_3$$

(184d)

in which

(184e) $k_1 = 1,(2)N_1 \;;\; k_2 = 0,(1),N_2 \;;\; k_3 = 0,(1),N_3 - 1$

Substitution into Eq. 183 and matching coefficients gives the following solution:

(185a) $$\bar{\underline{\theta}}_j(k_1,k_2,k_3) = \frac{\frac{1}{2}\bar{\underline{M}}_j^o(k_1,k_2,k_3)}{\sum_{i=1}^{3}(bK)_{ij}\,[\gamma_{ij} - \gamma_i(k_i)]} \quad \text{for } j = 1,2$$

in which

(185b) $$\gamma_i(k_i) = 1 - \cos\lambda_k^i$$

This completes the solution for the two three-dimensional sets of joint rotations in a regular flexural framework.

Probabilistic Analysis. – As has been demonstrated in previous lectures, use of discrete field methods for the static or dynamic analysis of a deterministically loaded totally specified structural system can be reduce the required computational effort by several orders of magnitude over that required for an open form simultaneous equation approach. Even more impressive, however, is the power of the closed form approach when such an analysis is not the end result but only an intermediate step in a more complex analysis such as:

1) a system optimization study;
2) a reliability analysis; or
3) a random vibrations analysis.

When additional analytic steps are required, the advantage of having the intermediate results in an analytic form rather than a set of numbers is very great indeed and may well be a necessary condition for the feasibility of carrying out the additional steps.

As an illustration of this point, consider the case of the shear frame with a random earthquake motion at the base (see Fig. 14). This frame was considered in Lecture VI with different damping and boundary conditions. The first step in the study is to obtain a dynamic analysis for an arbitrary ground motion G(t). Thus, with uniform story masses, ρ , and external damping, μ , the model is (see Eq. 96):

$$[\rho D_t^2 + \mu D_t - K\nabla_r]w(r,t) = 0 \quad (186a)$$

$$\Delta_r w(m,t) = 0, w(0,t) = G(t) \quad (186b,c)$$

The solution for this homogeneous equation with inhomogeneous boundary conditions can be expressed in terms of the series used previously (Eq. 97) plus a boundary function or as a series only, which is invalid at r = 0; i.e.,

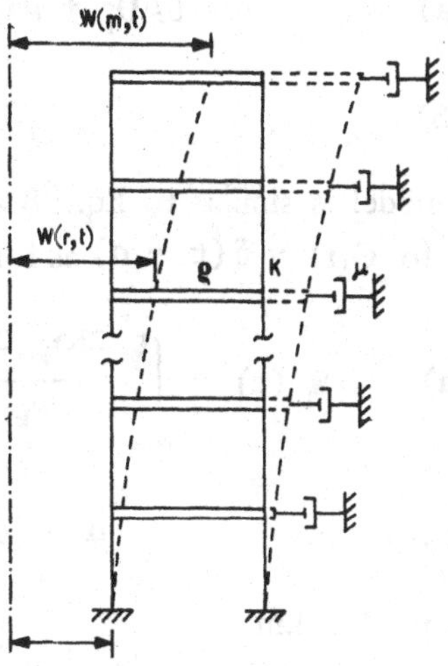

Fig. 14. Shear Frame with Ground Motion

$$w(r,t) = G(t) + \sum_{k=1}^{m} [W_k(t) - C_k G(t)] \sin \lambda_k r \qquad (187a)$$
$$r = 0,(1),m$$

$$\lambda_k = \frac{k - \frac{1}{2}}{m + \frac{1}{2}} \pi \qquad (187b)$$

or
$$w(r,t) = \sum_{k=1}^{m} W_k(t) \sin \lambda_k r \qquad r = 1,(1),m$$
(187c)

in which C_k is the expansion of unity in this series

(187d)
$$1 = \sum_{k=1}^{m} C_k \sin \lambda_k r$$

(187e)
$$C_k = \frac{\cot \frac{1}{2} \lambda_k}{m + \frac{1}{2}}$$

Substitution of Eq. 187a into 186a, expansion of the constant with respect to r, G(t), as a series (see Eq. 187d) and matching coefficients gives the following equation for $W_k(t)$:

(188a)
$$[\rho D_t^2 + \mu D_t + 2K\gamma_k] W_k(t) = 2K\gamma_k C_k G(t)$$

(188b)
$$\gamma_k = 1 - \cos \lambda_k$$

This model is similar to Eq. 98 and the solution can be written by integrating the result for $G(t) = \delta(t - \sigma)$ as follows:

(189a)
$$W_k(t) = \int_{-\infty}^{t} \frac{2K\gamma_k C_k}{\rho\beta_k} (e)^{-\alpha(t-\sigma)} \sin \beta_k (t-\sigma) G(\sigma) d\sigma$$

(189b,c)
$$\alpha = \frac{\mu}{2\rho}, \quad \beta_k^2 = \frac{2K}{\rho} \gamma_k - \alpha^2$$

or for r = 1,(1),m
$$w(r,t) = \frac{2K}{\rho\left(m + \frac{1}{2}\right)} \sum_{k=1}^{m} \frac{\sin \lambda_k}{\beta_k} \sin \lambda_k r \int_{-\infty}^{t} (e)^{-\alpha(t-\sigma)} \sin \beta_k (t-\sigma) G(\sigma) d\sigma$$
(189d)

For any given ground motion G(t), Eq. 189d can be used to find the response w(r,t); here, however, it is desired to use this closed form result to study the probabilistic characteristics of the response to a random ground excitation.(*)

The mean value of w(r,t) is found by taking the expected value of both sides of Eq. 189d. As the summation, integration, and ensemble averaging are all

(*) The author gratefully acknowledges the assistance of his colleague Dr. C.C. Tung, Associate Professor, Civil Engineering, North Carolina State University, in preparing this portion of the presentation.

linear operations, they are commutative and yield E[w(r,t)] by simply replacing $G(\sigma)$ in Eq. 189d by its expected value $E[G(\sigma)]$. For example, if the ground excitation has a mean value of zero, then $E[G(\sigma)] = 0$ resulting in $E[w(r,t)] = 0$.

For more useful information than the above mean values, the autocorrelation function (*) of w(r,t) (i.e., correlation between w(r,t) and $w(r,t+\tau)$) is required. The desired autocorrelation function is given by

$$E[w(r,t)w(r,t+\tau)] = \frac{4K^2}{\rho^2\left(m+\frac{1}{2}\right)^2}\sum_{k=1}^{m}\sum_{\ell=1}^{m}\frac{\sin\lambda_k \sin\lambda_\ell \sin\lambda_k r \sin\lambda_\ell r}{\beta_k\beta_\ell} \times$$

$$\int_{-\infty}^{t}\int_{-\infty}^{t+\tau}(e)^{-\alpha(t-\sigma)}(e)^{-\alpha(t+\tau-\sigma')}\sin\beta_k(t-\sigma)\sin\beta_\ell(t+\tau-\sigma') \times$$

$$E[G(\sigma)G(\sigma')]\,d\sigma\, d\sigma' \tag{190}$$

in which $E[G(\sigma)G(\sigma')]$ is the autocorrelation function of the ground motion G(t). The general case is tedious, but for the case where G(t) is an ideal white noise, the autocorrelation function is an impulse function permitting Eq. 190 to be readily evaluated; i.e., for $E[G(\sigma)G(\sigma')] = R_G\delta(\sigma'-\sigma)$

$$E[w(r,t)w(r,t+\tau)] = \frac{2K^2 R_G (e)^{-\alpha\tau}}{\rho^2\left(m+\frac{1}{2}\right)^2}\sum_{k=1}^{m}\sum_{\ell=1}^{m}\frac{\sin\lambda_k \sin\lambda_\ell \sin\lambda_k r \sin\lambda_\ell r}{\beta_k\beta_\ell} \times$$

$$[C_{k\ell}\cos\beta_k\tau + D_{k\ell}\sin\beta_k\tau] \tag{191a}$$

in which

$$C_{k\ell} = 8\,\alpha\,\beta_k\beta_\ell/F_{k\ell} \tag{191b}$$

$$D_{k\ell} = 2\,\beta_k[4\alpha^2 - \beta_k^2 + \beta_\ell^2]/F_{k\ell} \tag{191c}$$

$$F_{k\ell} = (4\alpha^2 - \beta_k^2 + \beta_\ell^2)^2 + 16\,\beta_\ell^2\alpha^2 \tag{191d}$$

It is noted that the expected value is independent of t implying that the response is stationary. The mean square values of w(r,t) (i.e., variances or squares of the standard deviations) are found by setting $\tau = 0$ in Eq. 191.

(*) Lin, Y.K., "Probabilistic Theory of Structural Dynamics, McGraw-Hill, Inc., 1967, pp. 37, 40-44, 95.

LECTURE XI

MULTIDIMENSIONAL SYSTEMS WITH DIFFICULT BOUNDARY CONDITIONS

Introduction. – This chapter is devoted to the analysis of a number of space structures with boundary conditions of a type commonly employed by designers but difficult to analyze by standard field methods. The basic concepts required to deal with these boundary conditions were covered in the course lectures but experience has shown that many students have difficulty applying them to multidimensional systems. The publishers were kind enough to allow the author these extra pages for the purpose of supplementing the lecture material with illustrative problems to better prepare the reader for the analysis of space lattices and mixed discrete-continuous systems with difficult boundary conditions.

Shear Grid on Four Rigid Supports. – For the first example of a structure with "difficult" boundary conditions, consider the uniform shear grid (as in Fig. 3 except boundary frames same as interior frames) resting on four symmetrically placed node supports. By referring to Eqs. 59-62, it is seen that the mathematical model and solution forms for supports at coordinates (a,b), (a,n-b), (m-a,b) ; and (m-a, n-b), are :

(192a) $$(K\Delta\!\!\!\!\nabla_r + \bar{K}\Delta\!\!\!\!\nabla_s)\, w(r,s) = -P(r,s)$$

(192b,c,d,e) $$\nabla_r w(0,s) = \Delta_r w(m,s) = \nabla_s w(r,0) = \Delta_s w(r,n) = 0$$

(192f,g,h,i) $$w(a,b) = w(a,n-b) = w(m-a,b) = w(m-a,n-b) = 0$$

(193a,b) $$\begin{bmatrix} P(r,s) \\ w(r,s) \end{bmatrix} = \sum_{k=0}^{m} \sum_{\ell=0}^{n} \begin{bmatrix} P_{k\ell} \\ W_{k\ell} \end{bmatrix} \cos\lambda_k \left(r + \frac{1}{2}\right) \cos\bar{\lambda}_\ell \left(s + \frac{1}{2}\right)$$

(193c,d,e) $$\lambda_k = \frac{k\pi}{m+1} \qquad \bar{\lambda}_\ell = \frac{\ell\pi}{n+1} \qquad W_{k\ell} = \frac{\frac{1}{2} P_{k\ell}}{K\gamma_k + \bar{K}\bar{\gamma}_\ell}$$

(193f,g) $$\gamma_k = 1 - \cos\lambda_k \qquad \bar{\gamma}_\ell = 1 - \cos\bar{\lambda}_\ell$$

It should be noted that the expressions for P(r, s) and w(r, s) cover the entire range r = 0,(1), m and s = 0, (1), n ; thus the concentrated reactions must be included in the loading term P(r, s) which is then self-equilibrating which in turn means that P_{oo} = 0 and W_{oo} is indeterminate. Each term of the solution satisfies the external boundary conditions, Eqs. 192 b,c,d,e and rigid body terms in the w(r, s) series are selected so as to satisfy the given support point displacements, Eqs. 192 f,g,h,i. For a specific loading consider a uniform applied node loading :

$$P(r,s) = P_o\left[1 - \frac{(m+1)(n+1)}{4}\left(\delta_r^a\delta_s^b + \delta_r^a\delta_s^{n-b} + \delta_r^{m-a}\delta_s^b + \delta_r^{m-a}\delta_s^{n-b}\right)\right] \tag{194a}$$

$$P_{k\ell} = 4P_o\omega_k\bar{\omega}_\ell\left[\delta_k^o\delta_\ell^o - \cos\lambda_k\left(a+\frac{1}{2}\right)\cos\bar{\lambda}_\ell\left(b+\frac{1}{2}\right)\right] \tag{194b}$$

$$k = 0,(2),m \qquad \ell = 0,(2),n \qquad \omega_k = 1 - \delta_k^o \qquad \bar{\omega}_\ell = 1 - \delta_\ell^o \tag{194c,d,e,f}$$

For the case of symmetry with respect to both $\frac{m}{2}$ and $\frac{n}{2}$, the rigid body term W_{oo} is found by substituting w(a, b) = 0 into the solution, Eq. 193 b,e. Numerical results for two specific grid configurations are shown in Tables 2 and 3 below* :

Table 2. – GRID DISPLACEMENTS FOR UNIFORM LOAD AND FOUR INSET REACTIONS

$(K/P_o)w(r,s)$ for case m = 4, n = 5, a = b = 1 and $K = 1.1\bar{K}$			
s \ r	0	1	2
0	2.4051	1.8583	2.0010
1	1.9065	0.0	1.2147
2	2.4051	1.8583	2.0010

(*) Dean, D.L., "Field Solutions for Shear Grids", *Journal of the Structural Division,* ASCE Proceedings Paper 8594, Vol. 97, No. ST12, Dec. 1971, pp. 2845-2860.

Table 3. – GRID DISPLACEMENTS FOR UNIFORM LOAD AND FOUR CORNER REACTIONS

$K/P_o w(r,s)$ for case $m = n = 12$, $a = b = 0$ and $K = \bar{K}$							
s \ r	0	1	2	3	4	5	6
0	0.0					symmetric about $r = s$	
1	20.625	28.494					
2	32.381	35.864	40.165				
3	39.653	41.416	43.965	46.490			
4	44.163	45.180	46.791	48.515	49.973		
5	46.657	47.350	48.504	49.800	50.942	51.713	
6	47.457	48.058	49.076	50.238	51.276	51.984	52.234

As a second example of a shear grid on for rigid supports, consider the structure shown in Fig. 3 to have finitely stiffened boundary frames at $s = 0$ and n and corner node supports. As in the first example, the physical boundary statement is that of zero external shear or free boundaries along the entire rectangular boundary ; however, in the first case the free boundary conditions could be satisfied identically by each term of a trigonometric series of functions which were symmetric about node lines one half module outside the actual structure. In this second system, however, the structure changes at $s = 0$ and n so that it is unlikely that we would be able to find a closed function that naturally satisfies the governing equation and the boundary conditions along the stiffened frames. Here the external boundary statements are replaced by the internal conditions which result from modification of the governing equation to reflect changes at the boundary, i.e. instead of Eqs. 192 b,c,d,e we have :

(195a) $$(K\Delta_r + \bar{K}\Delta\!\!\!\!\nabla_s)\, w(0,s) = -\, P(0,s)$$

(195b) $$(-K\nabla_r + \bar{K}\Delta\!\!\!\!\nabla_s)\, w(m,s) = -\, P(m,s)$$

$$(K^b \Delta\!\!\!/_r + \bar{K}\Delta_s)w(r,0) = -P(r,0) \tag{195c}$$

$$(K^b \Delta\!\!\!/_r - \bar{K}\nabla_s)w(r,n) = -P(r,n) \tag{195d}$$

in which K^b equals the stiffness of the boundary frames along s = 0 and n. Equations 195 a,b are equivalent to Eqs. 192 b,c. If $K^b = \infty$, the model would coincide with the system analyzed by Eqs. 60-62. If $K^b = K$ the model coincides with the above example and if $K^b = 1/2K$, the model can also be satisfied term by term by use of s functions of the form $\cos \ell\pi s/n$. For all other values, a closed form solution requires use of a particular solution plus a homogeneous solution whose terms are scaled so that the total solution satisfies the "unnatural" homogeneous boundary conditions. The particular solution will be selected to satisfy Eqs. 192 a,b,c,d,e or the system with regular free boundaries. The homogeneous solution will be written to satisfy the same model except that the loadings will consist only of external shear functions along s = 0 and n to duplicate the effects of the stiffened boundary frames, that is,

$$\bar{V}^h\left(r, \frac{-1}{n}\right) = \sum_{k=0}^{m} \left(\bar{V}_k^{a/s} \pm \bar{V}_k^{\$}\right) \cos \lambda_k\left(r + \frac{1}{2}\right) \tag{196}$$

For such a "loading" along s = 0 and n only, we require the following quantities :

$$\delta_s^0 \pm \delta_s^n = \sum_{\ell}^{n} \bar{C}_\ell \cos \bar{\lambda}_\ell\left(s + \frac{1}{2}\right) ; \begin{bmatrix} \ell = 0,(2),n & \text{for } + \\ \ell = 1,(2),n & \text{for } - \end{bmatrix} \tag{197a}$$

$$\bar{C}_\ell = \frac{4\bar{\omega}_\ell}{n+1} \cos \frac{1}{2}\bar{\lambda}_\ell \tag{197b}$$

The total solution can now be written as follows :

$$w(r,s) = \sum_{k=0}^{m}\sum_{\ell=0}^{n}\left(W_{k\ell}^p - \bar{V}_{k\ell}W_{k\ell}^h\right) \cos \lambda_k\left(r + \frac{1}{2}\right) \cos \bar{\lambda}_\ell\left(s + \frac{1}{2}\right) \tag{198a}$$

$$\bar{V}_{k\ell} = \begin{bmatrix} \bar{V}_k^{\$} & \text{if } \ell \text{ even} \\ \bar{V}_k^{a/s} & \text{if } \ell \text{ odd} \end{bmatrix} \tag{198b}$$

(198c,d) $$W^{p}_{k\ell} = \frac{\frac{1}{2} P_{k\ell}}{K\gamma_k + \bar{\bar{K\gamma}}_\ell} \quad ; \quad W^{h}_{k\ell} = \frac{\frac{1}{2} \bar{C}_\ell}{K\gamma_k + \bar{\bar{K\gamma}}_\ell}$$

All that remains is to find the values of the pseudo edge shears required to satisfy the actual homogeneous boundary conditions for the stiffened frame, Eqs. 195 c,d. Substitution of Eqs. 198 into these boundary conditions gives the following equation to be solved for $\bar{V}^{\$}_k$ and $\bar{V}^{a/s}_k$:

$$\left[1 + 2\gamma_k (K^b - K) \sum_{\ell}^{n} W^{h}_{k\ell} \cos \frac{1}{2} \bar{\lambda}_\ell\right] \begin{bmatrix} \bar{V}^{\$}_k \\ \bar{V}^{a/s}_k \end{bmatrix} = 2\gamma_k (K^b - K) \sum_{\ell}^{n} W^{p}_{k\ell} \cos \frac{1}{2} \bar{\lambda}_\ell$$

(199a,b)

in which $\ell = 0, (2), n$ to find $\bar{V}^{\$}_k$ and $\ell = 1, (2), n$ to find $\bar{V}^{a/s}_k$.

This completes the solution for the corner supported shear grid with edge stiffening except for the interdeterminacy of the $k = 0$, $\ell = 0$ terms which results from the fact that both the loading term P(r, s) and the boundary shear forces are self equilibrating, i.e. $P_{oo} = \bar{V}^{\$}_o = \bar{V}^{a/s}_o = 0$. It should also be noted that the summation on the left side of Eq. 199 is available as a single term hyperbolic function of γ_k . This term can be found alternatively as the boundary value of the Euler coefficient in a single series form of the homogeneous solution for a harmonic edge loading.

The recommended algorithm is to determine the rigid body term, $k = \ell = 0$, in Eq. 198 by use of the given value for the displacement at a support point, e.g. $w(a, b) = 0$; however, a better understanding of the nature of such field solutions is gained by formally solving for the rigid body term and substituting it back into Eq. 198 for a one-step explicit displacement formula. Thus, for the case of a uniform applied load, for which the loading and reaction coefficients are given by Eq. 194, the displacement field is :

$$w(r,s) = \sum_{\ell=0}^{n} \sum_{k=2}^{m} \frac{2\bar{\omega}_\ell F(r,s,a,b)}{K\gamma_k + \bar{\bar{K\gamma}}_\ell} \left[\cos \lambda_k \left(a + \frac{1}{2}\right) \cos \bar{\lambda}_\ell \left(b + \frac{1}{2}\right) P_o + \right.$$
$$\left. + \frac{\cos \frac{1}{2} \bar{\lambda}_\ell}{n+1} \bar{V}^{\$}_k \right] + P_o \sum_{\ell=2}^{n} \frac{\cos \bar{\lambda}_\ell \left(b + \frac{1}{2}\right)}{\bar{\bar{K\gamma}}_\ell} \left[\cos \bar{\lambda}_\ell \left(b + \frac{1}{2}\right) - \cos \bar{\lambda}_\ell \left(s + \frac{1}{2}\right)\right]$$

(200a)

$$F(r,s,a,b) = \cos \lambda_k \left(a + \frac{1}{2}\right) \cos \bar{\lambda}_\ell \left(b + \frac{1}{2}\right) - \cos \lambda_k \left(r + \frac{1}{2}\right) \cos \bar{\lambda}_\ell \left(s + \frac{1}{2}\right)$$

(200b)

in which $\ell = 0, (2), n$ and $k = 2, (2), m$ in the double summation and $\ell = 2, (2), n$ in the single summation and $\bar{V}_k^{\$}$ is found from Eq. 37a.

For a numerical illustration of the preceding solution for and edge stiffened shear grid, consider the case $m = 12$, $n = 9$, $\bar{K} = 3/2K$, $K^b = 4K$ and $a = b = 0$. The node displacements, as factors of P_o/K, are shown in Table 4 and the results of the intermediate calculations for the pseudo boundary shear forces are :

$$\frac{1}{P_o}\bar{V}_k^{\$} = (-4.892, -5.415, -4.935, -3.876, -2.460 \text{ and}, -.8421)$$

for k = 2,4,6,8,10, and 12 respectively.

TABLE 4 – DISPLACEMENTS FOR EDGE STIFFENED GRID WITH UNIFORM LOAD AND CORNER REACTIONS

$K/P_o\,w(r,s)$ for case $m = 12$, $n = 9$, $\bar{K} = 3/2\,K$, $K^b = 4K$ and $a = b = 0$							
s \ r	0	1	2	3	4	5	6
0	0.0	5.472	9.367	12.144	14.024	15.117	15.476
1	6.409	9.009	11.681	13.871	15.454	16.407	16.725
2	10.418	11.832	13.650	15.335	16.640	17.454	17.730
3	12.818	13.717	15.043	16.385	17.485	18.193	18.437
4	13.952	14.653	15.756	16.931	17.924	18.576	18.802

This completes the section on shear grids on four joint supports.

Trussed Stringer Bridge Deck. – As a second category of structures with "difficult" boundary conditions, we will analyze a torsionless bridge deck via a macro field approach, which is in contrast to the micro approach used in the preceding section on shear grids. The structure is similar to the grid analyzed in Lecture V except that here we will consider the case of flexible side supports and will also introduce the concept that one of the sets of grid members may be X-braced trusses instead of beams (see Fig. 15).

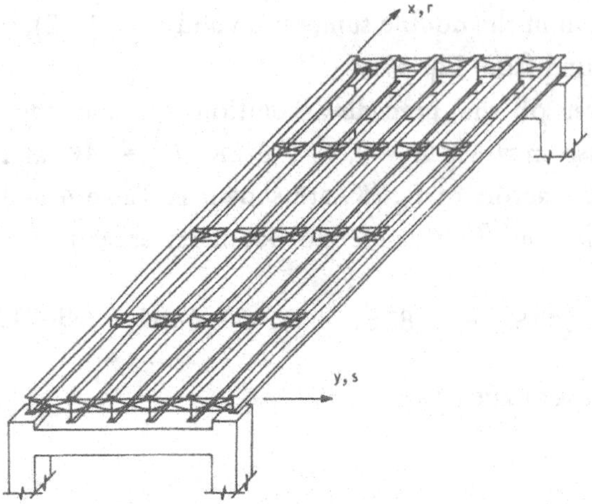

Fig. 15. Trussed Stringer Bridge Deck

Modification of the macro approach to satisfy "unnatural" homogeneous boundary conditions is similar to that used for the micro approach.

Namely, a scaled boundary function is added to the solution written for natural boundary conditions. The mathematical model for the micro analyses of a lattice consisted of a difference equation plus boundary conditions (e.g. see Eqs. 192). In the macro approach however, the model consists only of the governing summation equation with the statement of boundary conditions implicitly included in the kernel function or, as in this case, the kernel function plus a boundary function.

The unknowns for analysis of a torsionless grid with flexible side supports (with either beam or truss lateral distribution members) are the interactive node forces, R(r, s) between the stringers and diaphragms (see Eqs. 85) and the out-of-plane boundary node displacements along s = 0 and n.

(201) $$\bar{w}\left(r, {0 \atop n}\right) = \sum_{k=1}^{m-1} \left(U_k^{\$} \pm U_k^{a/s}\right) \sin \frac{k\pi r}{m}$$

The stringer displacement functions remain as given in Lecture V, i.e. Eqs. 82, 86a, 86e, 87a, 87f and 88. The diaphragm node displacements are :

(202a) $$\bar{w}\,(r,s) = \sum_{k=1} \left[U_k^{\$} + \left(1 - 2\frac{s}{n}\right) U_k^{a/s} \right] \sin \frac{k\pi r}{m} + \sum_{\beta=1}^{n-1} R(r,\beta)\bar{K}(s,\beta)$$

(202b) $$\bar{w}\,(r,s) = \sum_{k=1}^{m-1} \sum_{\ell=1}^{n-1} \left[U_{k\ell}\bar{C}_{\ell} + R_{k\ell}\bar{B}_{\ell} \right] \sin \frac{k\pi r}{m} \sin \frac{\ell\pi s}{n}$$

$$U_{k\ell} = \begin{bmatrix} U_k^{\$} & \text{if } \ell \text{ odd} \\ U_k^{a/s} & \text{if } \ell \text{ even} \end{bmatrix} \tag{202c}$$

$$\sum_{\ell}^{n-1} \bar{C}_\ell \sin \frac{\ell \pi s}{n} = \begin{bmatrix} 1 & \text{for } \ell \text{ odd} \\ 1 - 2\frac{s}{n} & \text{for } \ell \text{ even} \end{bmatrix} \qquad \bar{C}_\ell = \frac{2}{n} \cot \frac{\ell \pi}{2n} \tag{202d,e}$$

If the diaphragms are flexural members, $\bar{B}_\ell$ is given by Eqs. 87d,e. For trussed diaphragms, use the following formulas for the discrete kernel function coefficients(*):

$$\bar{B}_\ell = \frac{(\rho - 1)\bar{\gamma}_\ell + 2}{4h^2 Q_c \bar{\gamma}_\ell^2} \quad ; \quad \bar{\gamma}_\ell = 1 - \cos \frac{\ell \pi}{n} \tag{203a,b}$$

in which Q_c and Q_d equal area times elastic modulus divided by length cubed for the chord and diagonal members respectively and $\rho = Q_c/Q_d$.

The solution of this model for the interior interactive forces, i.e. R(r, s) for r = 1, (1), m – 1 and s = 1, (1), n – 1 or $R_{k\ell}$ for k = 1, (1), m – 1 and ℓ = 1, (1), n – 1, results from matching stringer and diaphragm interior node deflections. In other words, $w(\frac{L}{m}r,s) = \bar{w}(r,s)$. Matching coefficients in the series expression for these fields, Eqs. 87a,f and 202b gives :

$$R_{k\ell} = \frac{1}{B_k + \bar{B}_\ell} (W_{k\ell}^{o} - U_{k\ell} \bar{C}_\ell) \tag{204}$$

Thus, the displacement fields are now known except for the scale factor $U_k^{\$}$ and $U_k^{a/s}$ (or $U_{k\ell}$) in the homogeneous solution which appears as a part of $R_{k\ell}$ in Eq. 87a and in both terms in Eq. 202b. These coefficients are found by writing an expression for the boundary beam node deflections in terms of R(r, s) and any directly applied loads and then using Eq. 204 to eliminate the interactive force

(*) Dean, D.L., "Discussion of "Behavior of Howe, Pratt, and Warren Trusses" (by John D. Renton)", *Journal of the Structural Division*, ASCE, Sept. 1969, pp. 1997-2000.

coefficients. For example, consider the node deflections of the boundary beam at s = 0.

(205a) $$w\left(\frac{L}{m} r,0\right) = w^{o}(r,0) + \sum_{\beta=1}^{n-1} \sum_{\alpha=1}^{m-1} R(\alpha,\beta)\left(1 - \frac{\beta}{n}\right) K^{b}(r,\alpha)$$

(205b,c) $$\begin{bmatrix} w\left(\frac{L}{m} r,0\right) \\ w^{o}(r,0) \end{bmatrix} = \sum_{k=1}^{m-1} \begin{bmatrix} U_k \\ U_k^{o} \end{bmatrix} \sin \frac{k\pi r}{m}$$

(205d) $$U_k = U_k^{o} + \frac{n}{4} B_k^{b} \sum_{k=1}^{n-1} \bar{C}_{\ell} R_{k\ell}$$

in which U_k^{o} are coefficients of the series for the boundary beam node deflections due to loads, if any, applied directly to this member and $2/m\, B_k^{b}$ are the coefficients of the discrete kernel functions for the boundary member along s = 0, given by Eq. 74d,e for a flexural member except that B is replaced by B^{b}. Substitution of Eq. 204 into Eq. 205d and resolution into symmetric and antisymmetric components gives the following equations for $U_k^{\$}$ and $U_k^{a/s}$.

(206a,b) $$\left[1 + \frac{n}{4} B_k^{b} \sum_{\ell}^{n-1} \bar{F}_{k\ell} \bar{C}_{\ell}\right] \begin{bmatrix} U_k^{\$} \\ U_k^{a/s} \end{bmatrix} = \begin{bmatrix} U_k^{o\$} \\ U_k^{oa/s} \end{bmatrix} + \frac{n}{4} B_k^{b} \sum_{\ell}^{n-1} \bar{F}_{k\ell} W_{k\ell}^{o}$$

(206c) $$\bar{F}_{k\ell} = \frac{\bar{C}_{\ell}}{B_k + \bar{B}_{\ell}}$$

in which $\ell = 1$, (2), n – 1 for $U_k^{\$}$ and ℓ = 2, (2), n – 1 for $U_k^{a/s}$. $U_k^{o\$}$ and $U_k^{oa/s}$ are coefficients of the series for the boundary beam deflections due to the symmetries,1/2 q(x, 0)+1/2 q(x, n), and antisymmetric,1/2 q(x, 0)–1/2 q(x, n), components of the applied boundary beam loads respectively

(207) $$q\left(x, {}^{0}_{n}\right) = \sum_{i=1}^{\infty} \left(q_i^{*\$} \pm q_i^{*a/s}\right) \sin \frac{i\pi x}{L}$$

In order to numerically illustrate use of these formulas for the analysis of a beam-truss gridwork with bridge type boundaries, consider a system with the following physical properties : $L = 150$ ft.(45.72m); $\bar{L} = 30$ ft. (9.14 m); $m = 10$; $n = 6$; $B = 22.678 \times 10^{10}$ lb.in^2 (6.51×10^8 N-m^2); $B^b = 58.87 \times 10^{10}$ lb.in^2 (16.9×10^8 N-m^2); $a = 60$ in.(1.52 m); $h = 33$ in. (.838 m); $Q_c = 453.8$ lb/in^3 (1.23×10^8 N/m^3); $Q_{d'} = 180.64$ lb/in^3 ($.490 \times 10^8$ N/m^3); $\rho = 2.5122$

The loading on the interior beams will be taken as the first term expansion of a uniform deck loading equal to 64 lb/ft^2 (3.06×10^3 N/m^2); i.e., $\overset{*}{q}_{i\ell} = \overset{*}{q}_{11}\ \delta_i^1\ \delta_\ell^1$ in which $\overset{*}{q}_{11} = 42.24$ lb/in (7.40×10^3 N/m), from Eq.86e.

The boundary beam loads will also be taken as the first term expansion of their share of this uniform lane loading, i.e. $\overset{*}{q}{}_1^{a/s} = 0$; $\overset{*}{q}{}_i^{\$} = \overset{*}{q}_1\, \delta_i^1$, $\overset{*}{q}_1 = 16.98$ lb/ in (2.97×10^3 N/m) from Eq. 207.

Some results of intermediate calculations are $\overset{*}{B}_1 = 0.4752$ in^2/lb (6.89×10^{-5} m^2/N) from Eq. 69 e,f; $\overset{*b}{B}_1 = 0.1831$ in^2/lb (2.66×10^{-5} m^2/N), from Eq. 69 e,f with B replaced by B^b; $W_{k\ell} = \overset{*}{q}_{11}\ \overset{*}{B}_1\, \delta_k^1\, \delta_\ell^1$, $\overset{*}{q}_{11}\ \overset{*}{B}_1 = 20.07$ in.(51 cm) from Eq.74 b; $U_k^{o\$*} = \overset{*}{q}{}_1^{\$}\ \overset{*b}{B}_1\, \delta_k^1$, $\overset{*}{q}{}_1^{\$}\ \overset{*b}{B}_1 = 3.108$ in (7.9 cm), see Eq. 205 d; $B_1 = 2.640 \times 10^{-3}$in/lb ($1.51 \times 10^{-5}$m/N), from Eq. 74 c,d,e; $B_1^b = 1.017 \times 10^{-3}$ in/lb ($.581 \times 10^{-5}$ m/N), see Eq. 14; $\bar{B}_1 = 6.21 \times 10^{-5}$in/lb ($3.54 \times 10^{-7}$ m/N), from Eq. 203 a,b; $\bar{C}_1 = 1.244$, from Eq. 202 f; $U_1^{\$} = 8.857$ in.(.225 m) from Eq. 206 ; and $R_{11} = 3{,}350.5$ lb.(1.49×10^4 N), $R_{13} = -\ 1{,}117.5$ lb ($-.497 \times 10^4$ N), $R_{15} = -\ 229.6$ lb ($-.133 \times 10^4$ N) from eq. 204. The deflection and stress resultant fields were then found as follows : 1) Interior grid node deflections $s = 1,(1)n-1$ (in inches) $w(r,s) = [\,11.23 \sin \pi s/n + 2.95 \sin 3\pi s/n + .79 \sin 5\pi s/n\,] \sin \pi r/m$ fron Eq.87a or 12a; 2) Boundary node displacements (in inches)$w(r,\frac{o}{n}) = 8.857 \sin \pi r/m$ (from Eq. 205 b); 3) Interior continuous beam deflections (in inches) $w(x,s) = [\,11.23 \sin \pi s/n + 2.95 \sin 3\pi s/n + .79 \sin 5\pi s/n\,] \sin \pi x/L + 10^{-6}\ [\,68 \sin \pi s/n - 23 \sin 3\pi s/n - 6 \sin 5\pi s/n\,] \sin 19\pi x/L + \ \ldots$ (from Eq.86a); and 4)boundary continuous beam deflections (in inches)

$$w\left(x,\frac{o}{n}\right) = \left(\overset{*\$}{q_1} + \frac{m}{L}\frac{n}{4}\sum_{\ell=1,3}^{5} \bar{C}_\ell R_{1\ell}\right)\overset{*b}{B_1}\ \sin\frac{\pi x}{L}$$

$$- \left(\frac{mn}{4L}\sum_{\ell=1,3}^{5} \bar{C}_\ell R_{1\ell}\right)\overset{*b}{B_{19}}\ \sin\frac{19\pi x}{L} + \dots$$

$$= 8.857 \sin\frac{\pi x}{L} - 44 \times 10^{-6} \sin\frac{19\pi x}{L} + 30 \times 10^{-6} \sin\frac{21\pi x}{L} + \dots$$

The General Boundary Case. – The most general basic boundary condition for a grid is that system with flexible supports on all four sides and zero corner displacements (see Fig. 16). The variation for which the reactions are located at other than the corner nodes is dealt with by including the reactions in the general load term. This results in a self equilibrating load and zero corner reaction as the basic solution which requires an additional rigid body movement to correct the displacement field so as to give zero deflection at the reaction points rather than at the grid corners. For this system one requires in addition to Eq. 205d, an expression for the boundary node series coefficients along the transverse boundary component, i.e. :

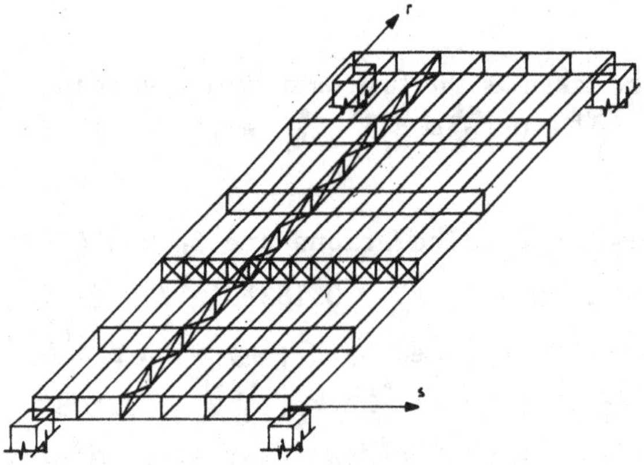

Fig. 16. Corner Supported Gridwork

(208) $$V_\ell = V_\ell^{o} - \frac{m}{4}\bar{B}_\ell^{b}\sum_{k=1}^{m-1} C_k R_{k\ell}$$

in which the new terms are inductively defined by reference to the definitions following Eq. 205d. For displacements along r = 0 and m, the expressions for the stringer deflections, Eqs. 82, 86 and 87, must be modified to include rigid body boundary functions as shown in Eqs. 202 for the diaphragm deflections ; i.e.

(209a) $$w\left(\frac{o}{m},s\right) = \sum_{\ell=1}^{n-1}\left(V_\ell^{\$} \pm V_\ell^{a/s}\right) \sin\frac{\ell\pi s}{n}$$

$$w\left(\frac{L}{m}r,s\right) = \sum_{k=1}^{m-1}\sum_{\ell=1}^{n-1}(W^q_{k\ell} + V_{k\ell}C_k - R_{k\ell}B_k)\sin\frac{k\pi r}{m}\sin\frac{\ell\pi s}{n} \tag{209b}$$

Substitution of these modified stringer equations into the node compatibility equation, $w(L/m\,r,s) = \bar{w}(r,s)$ gives the following result for the interactive force coefficients in terms of the applied load coefficients and the unknown boundary displacement coefficients :

$$R_{k\ell} = \frac{1}{B_k + \bar{B}_\ell}(W^o_{k\ell} + V_{k\ell}C_k - U_{k\ell}\bar{C}_\ell) \tag{210}$$

Substitution of Eq. 210 for $R_{k\ell}$ into Eqs. 205d and 208 gives the following equations to be solved for $U^\$_k$ and $V^\$_\ell$:

$$\left[\frac{4}{n}\frac{1}{B^b_k} + \sum_{\ell=1,3\ldots}^{n-1}\bar{C}_\ell\bar{F}_{k\ell}\right]U^\$_k - C_k\sum_{\ell=1,3\ldots}^{n-1}\bar{F}_{k\ell}V^\$_\ell = \frac{4}{n}\frac{1}{B^b_k}U^{o\$}_k + \sum_{\ell=1,3\ldots}^{n-1}\bar{F}_{k\ell}(W^q_{k\ell} - \bar{W}^q_{k\ell}) \tag{211a}$$

$$-\bar{C}_\ell\sum_{k=1,3\ldots}^{m-1}F_{k\ell}U^\$_k + \left[\frac{4}{m}\frac{1}{\bar{B}^b_\ell} + \sum_{k=1,3\ldots}^{m-1}C_kF_{k\ell}\right]V^\$_\ell = \frac{4}{m}\frac{1}{\bar{B}^b_k}V^{o\$}_\ell - \sum_{k=1,3}^{m-1}F_{k\ell}(W^q_{k\ell} - \bar{W}^q_{k\ell}) \tag{211b}$$

in which $F_{k\ell}\ \dfrac{C_k}{B_{k\ell} + \bar{B}_\ell}$

(Analogous expressions for the antisymmetric and mixed symmetric-antisymmetric terms may be written by induction as illustrated by Eqs. 206a,b. Solutions of these simultaneous summation equations for $U^\$_k$ and $V^\$_\ell$ will complete the analysis of the object grid system with general boundary conditions.

A direct simultaneous field solution of Eqs. 211a and 211b, i.e., a formula containing m and n as parameters, is beyond the present state of the art of discrete field mechanics. However, the interactive effects of $V^\$_\ell$ in Eq. 211a and $U^\$_k$ in Eq. 211b are small engough so that a cyclic approach converges rapidly to very accurate results which can then be used to complete exact closed expressions for the desired deflection and stress fields. The recommended procedure is to :

1) assume the smaller set, say $\overset{\$}{V}_{\ell}$, equal to zero and solve Eq. 211a for an initial set of values for $\overset{\$}{U}_{k}$;

2) Substitute these values for $\overset{\$}{U}_{k}$ into Eq. 211b ans solve for a better initial set of values for $\overset{\$}{V}_{\ell}$;

3) Substitute these values for $\overset{\$}{V}_{\ell}$ back into Eq. 211a to find corrected values for $\overset{\$}{U}_{k}$, etc. until the values for $\overset{\$}{U}_{k}$ and $\overset{\$}{V}_{\ell}$ have converged to the desired level of accuracy.

For special cases this algorithm can be improved. For example, consider the loading to be harmonic so that $\overset{o}{W}_{k\ell} = \overset{o}{W}_{11}\, \delta_k^1 \delta_\ell^1$. In this case an improvement results from changing the first step as follows : Solve Eqs. 211a and 211b simultaneously for $(\overset{\$}{U}_{1})^{(0)}$ and $(\overset{\$}{V}_{1})^{(0)}$ by assuming all other coefficients zero. One can then proceed as above to find $(\overset{\$}{U}_{k})^{(1)}$ $(\overset{\$}{V}_{\ell})^{(1)}$ etc. This completes the algorithm for the analysis of the object problem.

The Ribbed Plate Bridge Deck. – As a third category of structures with "difficult" boundary conditions, we will consider a mixed discrete-continuous system. Specifically, the object of this section is the analysis of a metal bridge deck consisting of a continuous thin element top membrane that is in continuous contact with and composite with a set of ribs parallel to the direction of traffic. All elements are proportioned and detailed so that there is significant "T-beam" or composite action between the stringers and the plate (or membrane) but the plate itself has negligible resistance to out-of-plane deformations. That is, the plate behaves as a membrane which contributes to the effective upper flange area of the stringers by resisting in-plane deformations, but does not have sufficient independent flexural rigidity to distribute deck loading to the stringers through transverse shear action. A bridge deck comprised of a thin steel plate welded to closely spaced stringers (see Fig. 17) is an example of such a system. This system is similar to the ribbed plate dealt with in Lecture VII except that here there are flexible supports along $x = 0$ and a (or $r = 0$ and m). Thus, the kernel functions for the simply supported membrane, Eqs. A-4.5 through A-4.18, will not be sufficient to express the plate deflections. The membrane load effects will be determined, analogous to a particular solution, by use of the simple support kernel functions which were derived using the model A-4.1 through A-4.4 plus the boundary conditions $v(\substack{o\\a},y) = n_x(\substack{o\\a},y) = 0$. This particular solution will then be corrected by the addition of a homogeneous solution

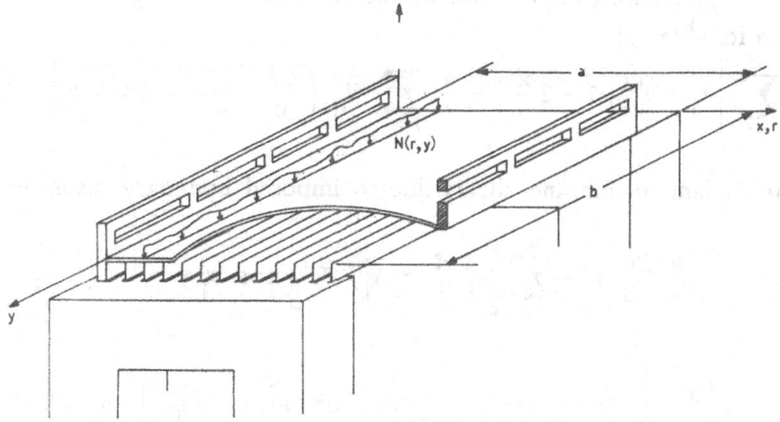

Fig. 17. Composite Membrane Ribbed Plate Model

which satisfies the homogeneous version of the model A-4.1 through A-4.4 plus the boundary conditions $n_x^h\left({}^{o}_{a},y\right) = 0$ and:

$$v^h\left({}^{o}_{a},y\right) = \sum_{j=0}^{\infty} (V_j^{\$} \pm V_j^{a/s}) \cos \bar{\alpha}_j y \tag{212a}$$

$$V_{ij} = \begin{bmatrix} V_j^{\$} & \text{for } i \text{ odd} \\ V_j^{a/s} & \text{for } i \text{ even} \end{bmatrix} \tag{212b}$$

$$v^h(x,y) = \frac{4}{a} \sum_{j=0}^{\infty} \sum_{i=1}^{\infty} \bar{V}_{ij} \overset{*}{\bar{B}}_{ij} \sin \alpha_i x \cos \bar{\alpha}_j y \tag{212c}$$

$$u^h(x,y) = \frac{4}{a} \sum_{j=1}^{\infty} \sum_{i=0}^{\infty} \overset{*}{\phi}_i \bar{V}_{ij} \overset{*}{\bar{A}}_{ij} \cos \alpha_i x \sin \bar{\alpha}_j y \tag{213a}$$

$$\overset{*}{\bar{A}}_{ij} = \frac{\bar{\alpha}_j (\bar{\alpha}_j^2 - \mu\alpha_i^2)}{(\alpha_i^2 + \bar{\alpha}_j^2)^2} \tag{213b}$$

Many applications require use of the following more rapidly convergent mixed formula for $v^h(x, y)$:

$$v^h(x,y) = \sum_{j=0}^{\infty} \left[V_j^{\$} + V_j^{a/s}\left(1 - 2\frac{x}{a}\right) + \frac{4}{a}\sum_{i=1}^{\infty} \bar{V}_{ij}\left(\bar{B}^{*}_{ij} - \frac{1}{\alpha_i}\right)\sin \alpha_i x\right]\cos \alpha_j y$$

(214)

The inplane membrane shears due to imposed boundary displacements are :

(215a) $$n^h_{xy}\left({}^{o}_{a}, y\right) = \sum_{j=1}^{\infty}\left[V_j^{\$}\bar{T}_j^{\$} \pm V_j^{a/s}\bar{T}_j^{a/s}\right]\cos \alpha_j y$$

(215b) $$\begin{bmatrix} T_j^{\$} \\ T_j^{a/s} \end{bmatrix} = -\frac{K}{2}(1-\mu^2)\bar{\alpha}_j\left[\frac{\sinh a\bar{\alpha}_j \pm a\bar{\alpha}_j}{\cosh a\bar{\alpha}_j \pm 1}\right]$$

In addition to the particular rib line solutions given by Eqs. A-4.12 through A-4.18, we will need to refer to the following homogeneous rib line descriptors :(*)

(216a) $$u^h(r,y) = \frac{4}{m}\sum_{j=1}^{\infty}\sum_{k=0}^{m}\bar{V}_{kj}\,\bar{A}_{kj}\cos\frac{k\pi r}{m}\sin\bar{\alpha}_j y$$

(216b) $$v^h(r,y) = \frac{4}{m}\sum_{j=0}^{\infty}\sum_{k=1}^{m-1}\bar{V}_{kj}\,\bar{B}_{kj}\sin\frac{k\pi r}{m}\cos\bar{\alpha}_j y$$

$$\bar{A}_{kj} = \frac{\Phi_k}{4\bar{D}_{kj}}\left[(1-\mu)\sinh\lambda_j - \frac{(1+\mu)\lambda_j}{\bar{D}_{kj}}\left(1 - \cosh\lambda_j\cos\frac{k\pi}{m}\right)\right]$$

(216c)

The interior loading function is expressed in series form as given by Eq. 110a in which the coefficients are found from :

(*) Dean, D.L. and Abdel-Malek, R.A., "Rational Analysis of Orthotropic Bridge Decks", *International Journal of Mechanical Sciences,* Pergamon Press, Vol. 16, No. 3, March, 1974, pp. 173-192.

$$N_{kj} = \frac{4}{mb} \int_o^b \sum_{r-1}^{m-1} N(r,y) \sin \frac{k\pi r}{m} \sin \bar{\alpha}_j y \, dy \tag{217}$$

The unknown interior rib-plate interactive forces y(r, y) are also expressed in series form by Eq. 110b.

The in-plane plate displacements along the rib lines in the x and y directions are determined as follows :

$$u(r,y) = u^h(r,y) + \sum_{\alpha=1}^{m-1} \int_o^b Y(\alpha,\eta) K^{uy}(r,y,\alpha,\eta) d\eta \tag{218}$$

$$v(r,y) = v^h(r,y) + \sum_{\alpha=1}^{m-1} \int_o^b Y(\alpha,\eta) K^{vy}(r,y,\alpha,\eta) d\eta \tag{219}$$

in which u^h, K^{uy}, v^h, and K^{vy} are given respectively by Eqs. 216a, A-4.12, 216b, and A-4.13.

$$v_R(r,y) = \int_o^b [N(r,y) K^{vz}(y,\eta) + Y(r,y) K^{vy}(y,\eta)] d\eta \tag{220}$$

$$w_R(r,y) = \int_o^b [N(r,y) K^{wz}(y,\eta) + Y(r,y) K^{wy}(y,\eta)] d\eta \tag{221}$$

in which K^{vz}, K^{vy}, K^{wz} and K^{wy} are given by Eqs. A-4.22 and A-4.23.

The series expressions for these four deflection functions can be written by use of standard orthogonality relations as follows :

$$u(r,y) = \sum_{j=1}^{\infty} \sum_{k=0}^{m} \left[\frac{4}{m} \bar{V}_{kj} \bar{A}_{kj} + Y_{kj} A_{kj} \right] \cos \frac{k\pi r}{m} \sin \bar{\alpha}_j y \tag{222}$$

$$v(r,y) = \sum_{j=1}^{\infty} \sum_{k=1}^{m-1} \left[\frac{4}{m} \bar{V}_{kj} \bar{B}_{kj} + Y_{kj} B_{kj} \right] \sin \frac{k\pi r}{m} \cos \bar{\alpha}_j y \tag{223}$$

$$v_R(r,y) = \sum_{j=1}^{\infty} \sum_{k=1}^{m-1} [N_{kj} D_j^* - Y_{kj} B_j^*] \sin \frac{k\pi r}{m} \cos \bar{\alpha}_j y \tag{224}$$

$$w_R(r,y) = \sum_{j=1}^{\infty} \sum_{k=1}^{m-1} [N_{kj} A_j^* - Y_{kj} D_j^*] \sin \frac{k\pi r}{m} \sin \bar{\alpha}_j y \tag{225}$$

in which $\bar{V}_{kj}$, $\bar{A}_{kj}$, A_{kj} , $\bar{B}_{kj}$, B_{kj} , $\overset{*}{D}_j$, $\overset{*}{B}_j$ and $\overset{*}{A}_j$ are given by Eqs. 212b, 216c, A-4.14, A-4.18, A-4.15 and A-4.23.

The solution for the case of known side boundary deflections (i.e. $\bar{V}_{kj}$, given) is obtained by expressing compatibility of longitudinal deflection between the plate and rib tops, $v(r, y) = v_R(r, y)$, and solving for the interactive force coefficients. The result is :

(226)
$$Y_{kj} = \frac{\overset{*}{D}_j N_{kj} - \frac{4}{m} \bar{B}_{kj} \bar{V}_{kj}}{B_{kj} + \overset{*}{B}_j}$$

The side boundary deflections for the system with flexible side supports are found by matching the longitudinal plate boundary deflection with the corresponding deflections along the top of the boundary ribs. The total loading on the boundary ribs includes applied loads plus membrane shears on the plate boundary due to the interactive forces and the boundary displacements. Matching coefficients gives the following results for the boundary displacement coefficients :

(227)
$$N\binom{0}{m}\, y) = \sum_{j=1}^{\infty} (P_j^{\$} \pm P_j^{a/s}) \sin \bar{\alpha}_j y$$

(228a)
$$V_j^{\$} = \frac{\overset{*b}{D}_j P_j^{\$} + \overset{*b}{B}_j \overset{*}{D}_j T_j^{N\$}}{1 - \overset{*b}{B}_j \bar{T}_j^{\$} + \overset{*b}{B}_j \bar{T}_j^{y\$}}$$

(228b)
$$V_j^{a/s} = \frac{\overset{*b}{D}_j P_j^{a/s} + \overset{*b}{B}_j \overset{*}{D}_j T_j^{Na/s}}{1 - \overset{*b}{B}_j \bar{T}_j^{a/s} + \overset{*b}{B}_j \bar{T}_j^{ya/s}}$$

in which $\overset{*b}{D}_j$ and $\overset{*b}{B}_j$ are the quantities defined in Eq. A-4.23 with B replaced by B^b, the boundary rib rigidity ; $\overset{*}{D}_j$ is as defined in Eq. A-4.23 using rigidity of interior ribs ; $\bar{T}^{\$}$, and $\bar{T}_j^{a/s}$ are the coefficients of boundary membrane shear due to the boundary displacements (see Eqs. 215b) ; and $T_j^{N\$}$, and $T_j^{Na/s}$ and $\bar{T}_j^{y\$}$ and $\bar{T}_j^{y\,a/s}$ are coefficients of the boundary membrane shear due to the interactive forces Y(r, y). $T_j^{N\$}$ and $T_j^{Na/s}$ represent the components resulting from the out-of-plane loading and $\bar{T}_j^{y\$}$ and $\bar{T}_j^{y\,a/s}$ represent the components resulting from boundary displacements. These coefficients are found by use of the formulas below :

$$\begin{Bmatrix} T_j^{N\$} \\ \\ T_j^{Na/s} \end{Bmatrix} = \sum_k \frac{\bar{B}_{kj} N_{kj}}{B_{kj} + \overset{*}{B}_j} \quad \text{for} \quad \begin{Bmatrix} k = 1,(2), m-1 \\ \\ k = 2,(2), m-1 \end{Bmatrix} \tag{229a,b}$$

$$\begin{Bmatrix} \bar{T}_j^{y\$} \\ \\ T_j^{ya/s} \end{Bmatrix} = \frac{4}{m} \sum_k \frac{(\bar{B}_{kj})^2}{B_{kj} + \overset{*}{B}_j} \quad \text{for} \quad \begin{Bmatrix} k = 1,(2), m-1 \\ \\ k = 2,(2), m-1 \end{Bmatrix} \tag{230a,b}$$

This completes the rib line formulas for a composite membrane analysis of the orthotropic deck with flexible side supports.

The Continuous Deflection Fields can be found by use of the kernel function for continuous coordinates, Eqs. A-4.5, A-4.6. The results are :

$$u(x,y) = u^h(x,y) + \sum_{j=1}^{\infty} \sum_{i=1}^{\infty} \frac{m}{a} Y_{ij} \overset{*}{A}_{ij} \cos \alpha_i x \sin \bar{\alpha}_j y \tag{231a}$$

$$v(x,y) = v^h(x,y) + \sum_{j=1}^{\infty} \sum_{i=1}^{\infty} \frac{m}{a} Y_{ij} \overset{*}{B}_{ij} \sin \alpha_i x \cos \bar{\alpha}_j y \tag{231b}$$

in which the homogeneous solutions u^h and v^h are given by Eqs. 213, 212c and A-4.11 respectively ; the coefficients $\overset{*}{A}_{ij}$ and $\overset{*}{B}_{ij}$ are given by Eqs. A-4.8 and A-4.9 and the interactive force coefficients Y_{ij} are given by Eq. 28 where the quantity is sinewise cyclic on the first index, i.e. $Y_{21\,m-k,j} = -Y_{21\,m-k,j} = = Y_{kj}$.

In order to illustrate use of the formulas for a composite membrane analysis of a metal ribbed plate bridge deck, consider a structure with input data as follows : a = 144 in., b = 72 in., m = 12, t = 0.5 in., E = 29,000ksi, μ = 0.3, B = 5/3 E kip-in^2, e = 2 in.,ρ^2 = 4/3in^2. The rib line loading is symmetric with respect to y = b/2 and antisymmetric with respect to x = a/2, i.e. N_{21} = .02kip/in, other N_{kj} = 0. Some of the intermediate results are : $\overset{*}{B}_1$ = 0.05796 in.^{2}/kip, $\overset{*}{D}_1$ = 0.49812 in^2/kip, and $\overset{*}{A}_1$ = 5.7080 in.^{2} /kip (Eq. A-4.23) ; K_* = 15,934 kip/in, (Eq. A-4.2) ; $\overset{*}{B}_{2,1}$ = 0.03179 in.^{3} /kip, $B_{22,1}$ = 0.00077 in.^{3} /kip, $B_{26,1}$ = 0.00055 in.^{3} /kip (Eq. A-4.8); $B_{2,1}$ = 0.002829 in.^{2} /kip (Eq. A-4.15) or $B_{2,1}$ ≈ 0.00265 in^2 /kip (Eq. A-4.16). The following additional data on the side boundary ribs is required : B^b = 45/8 E kip-in^2 ; e^b = 3 in., $(\rho^b)^2$ = 3 in.^{2} and $P_j^{\$} = P_j^{a/s}$ = 0.

Some additional intermediate results, are :

$\bar{B}_{21} = 0.29177$ (Eq.A-4.18); $T_1^{Na/s} = 0.0960$ kip^2/in.^{3} (Eq.229b);

$\bar{T}_1^{y\,a/s} = 2.90$ ksi (Eq.230b); $\bar{T}_1^{a/s} = -310$ ksi(Eq.215b); $\overset{*}{B}{}_1^{b} = 0.0386$ in.^{2}/kip

(Eq.A-4.23 modified with B^b); $V_1^{a/s} = 1.411 \times 10^{-4}$ in.(Eq.228b);

$Y_{21} = 0.16366$ kip/in., $Y_{41} = 3.8 \times 10^{-4}$ kip/in., $Y_{61} = -2.9 \times 10^{-4}$ kip/in.,

$Y_{81} = -1.9 \times 10^{-4}$ kip/in. and $Y_{10,1} = .9 \times 10^{-4}$ kip/in.(Eq.226); and

$\overset{*}{\bar{B}}_{21} - \dfrac{1}{\alpha_1} = -18.91$ and $\overset{*}{\bar{B}}_{61} - \dfrac{1}{\alpha_1} = -1.66$ (Eq.A-4.11).

The field of interior rib line deflections in the y direction (inch units) is found by substituting into Eq. 223 or 224. For example, at $r = 3$ the result is $v(3, y) = 4.65 \times 10^{-4} \cos \pi y/b$ (or by a one term approximation $4.77 \times 10^{-4} \cos \pi y/b$). The in-plane boundary deflection, as defined by Eq. 212a is $v(0, y) = 1.411 \times 10^{-4} \cos \pi y/b$. The field of interior rib line deflections in the z direction (inch units) is given by Eq. 225. At $r = 3$, the results is $w(3, y) = 0.0325 \sin \pi y/b$ (or $.0326 \sin \pi y/b$ with a one term approximation). The deflection in the z or out-of-plane direction along the boundary rib can be found in terms of the in-plane deflection by elementary beam theory ; i.e. from Eq. A-4.19.

$$W_j^{a/s} = \frac{e^b V_j^{a/s}}{(\rho^2 + e^2)^b \bar{\alpha}_j} \tag{232}$$

$$\text{or} \quad w(o,y) = 8.1 \times 10^{-4} \sin \frac{\pi y}{b} .$$

The continuous in-plane deflection fields are given by Eqs. 231a,b. The accuracy available through use of a truncated infinite series is shown by use of the solution for $v(x, y)$ (using $i = 2, 6, 10$ for the homogeneous component and $i = 2, 22, 26$ – terms containing Y_{21} – for the inhomogeneous component) to find displacements along a rib line, which can be compared with exact expressions found by use of a finite series. The result is $v(a/4,y) = 4.53 \times 10^{-4} \cos \pi y/b$, which should be compared with $v(3, y)$ above.

The Plate-Stringer-Diaphragm Metal Bridge Deck. – As a fourth structure with "difficult" boundary conditions, we will consider a mixed discrete-continuous system in the category of two dimensional lattices with two dimensional elements.

Specifically, the object problem is that of analyzing a bridge deck consisting of a thin deck monolithic with a set of stringers which, in turn, are braced by a set of diaphragms. There is not composite action between the deck plate and the diaphragms. (see Fig. 18). The analysis is similar to that of the previous section

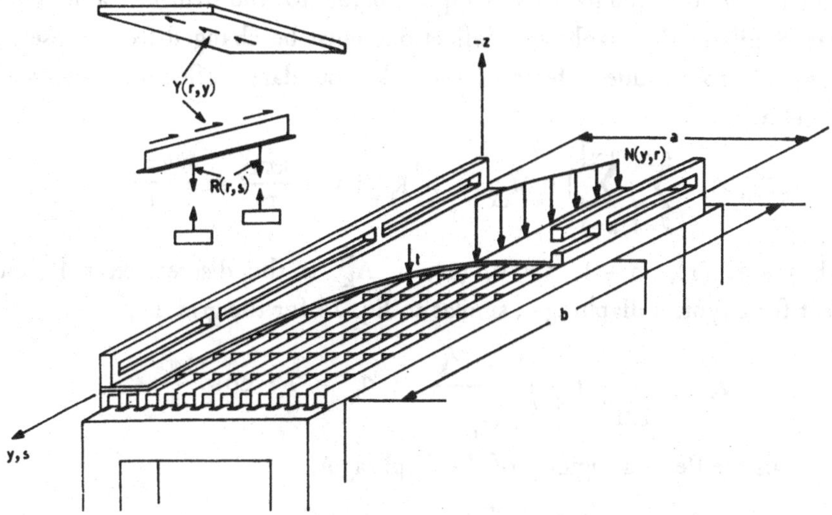

Fig. 18. Thin Element Plate-Stringer-Diaphragm Deck

except for the addition of the diaphragms and similar to the grid-plate structure considered in Lecture VII except that here we have the complication of satisfying unnatural homogeneous side boundary conditions.

The unknowns here are : 1) the discrete-continuous interactive forces between the deck and the stringers, Y(r, s) (see Eq. 110b) ; 2) the discrete interactive forces between the stringers and diaphragms, R(r, s) (see Eq. 85a) ; 3) the continuous in-plane displacements at the top of the boundary stringers, $v^h\left({}^{o}_{a},y\right)$ (see Eq. 212a); and 4) the continuous out-of-plane displacements of the boundary beams, $w\left({}^{o}_{a},y\right)$ (see Eq. 233 below).

$$w\begin{pmatrix}0\\a\end{pmatrix}, y) = \sum_{j=1}^{\infty} (W_j^{\$} \pm W_j^{a/s}) \sin \bar{\alpha}_j y \tag{233a}$$

$$\bar{W}_{kj} = \begin{bmatrix} W_j^{\$} & \text{for } k \text{ odd} \\ W_j^{a/s} & \text{for } k \text{ even} \end{bmatrix} \tag{233b}$$

The loading is as given in the previous section, Eqs. 110a and 227.

The continuous and rib-line deck displacements here are unchanged from the previous sections, i.e. see Eqs. 212, 213, 214, 216, 218, 219, 222, 223 and 231. The interior stringer or rib displacements, however, are changed to include the effects of the interactive forces between stringers and diaphragms, i.e. see Eqs. 117a,b for continuous quantities and Eqs. 120a,b,c for the stringer-diaphragm node quantities. Similarly the diaphragm deflections must be changed from those shown in Lecture VII to include effects of the side boundary deflections, i.e. see Eqs. 234-237 below :

$$(234) \qquad w(r,s) = \sum_{\ell=1}^{n-1}\sum_{k=1}^{m-1}[C_k \bar{W}^f_{k\ell} + A^d_k R_{k\ell}] \sin\frac{k\pi r}{m} \sin\frac{\ell\pi s}{n}$$

in which $r = 1, (1), m-1$, $s = 0, (1), n$, A^d_k is the discrete kernel function coefficient for a typical diaphragm, similar to Eq. 74 for the grid, i.e.

$$(235a,b) \qquad A^d_k = \frac{1}{12B^d}\left(\frac{a}{m}\right)^3 \frac{3-\sigma_k}{(\sigma_k)^2} \quad ; \quad \sigma_k = 1-\cos\frac{k\pi}{m}$$

and B^d equals the flexural rigidity of the diaphragm,

$$(236) \qquad \bar{W}^f_{k\ell} = \sum_{J=-\infty}^{+\infty} \bar{W}_{k,\,2Jn+\ell}$$

$$(237a,b) \qquad 1 = \sum_{k=1,3,\dots}^{m-1} C_k \sin\frac{k\pi r}{m} \qquad\qquad 1-2\frac{r}{m} = \sum_{k=2,4,\dots}^{m-1} C_k \sin\frac{k\pi r}{m}$$

$$(237c) \qquad C_k = \frac{2}{m}\cot\frac{k\pi}{2m}$$

The equations necessary to solve for the unknown coefficients Y_{kj} , $R_{k\ell}$, $V^{\$}_j$ or $V^{a/s}_j$ and $W^{\$}_j$ or $W^{a/s}_j$ result from matching in-plane stringer-plate displacements, out-of-plane stringer-diaphragm displacements, in-plane boundary-stringer displacements and out-of-plane boundary stringer displacements. The effects of the diaphragm end reactions on the boundary stringers are found as with the trussed tringer bridge deck, i.e. see Eqs. 205. The equations to be solved simultaneoulsy for these unknown series coefficients are as follows (shown for symmetric component of boundary displacements) :

$$(B_{kj} + \overset{*}{B}_j)\, Y_{kj} + \frac{n}{b}\, \overset{*}{D}_j R_{kj} + \frac{4}{m}\, \bar{B}_{kj} V_j^{\$} = \overset{*}{D}_j N_{kj} \qquad (238)$$

$$\sum_J \overset{*}{D}_{2Jn+\ell}\, Y_{k,\,2Jn+\ell} + (A_k^d + A_\ell) R_{k\ell} + C_k \sum_J W^{\$}_{2Jn+\ell} = \sum_J \overset{*}{A}_{2Jn+\ell}\, N_{k,\,2Jn+\ell} \qquad (239)$$

$$\sum_k \left[\overset{*b}{B}_j \bar{B}_{kj} Y_{kj} + \frac{mn}{4b} \overset{*b}{D}_j C_k R_{kj} \right] + \left[\overset{*b}{B}_j \bar{T}_j^{\$} - 1 \right] V_j^{\$} = - \overset{*b}{D}_j P_j^{\$} \qquad (240)$$

$$\sum_k \left[\overset{*b}{D}_j \bar{B}_{kj} Y_{kj} + \frac{mn}{4b} \overset{*b}{A}_j C_k R_{kj} \right] + \overset{*b}{D}_j \bar{T}_j^{\$} V_j^{\$} - W_j^{\$} = - \overset{*b}{A}_j P_j^{\$} \qquad (241)$$

in which $J = -\infty\ (1), +\infty$ with convergence about $J = 0$ and $k = 1, (2), m - 1$. To solve for antisymmetric boundary displacements replace all \$ quantities by the analogous a/s quantities and use $k = 2, (2), m - 1$. It is apparent that this model cannot be dealt with as a set of algebraic equations due to inconformability – i.e. Eq. 238 is for kj indexed quantities, Eq. 239 is for $k\ell$ indexed quantities and Eqs. 240 and 241 are for j indexed quantities – and the fact that some terms are sums ; however, the simultaneous equations can be solved by successive elimination of unknowns.

Before proceedings to describe the algorithm to solve the above model, Eqs. 238-241, for a plate stringer diaphragm system, it should be noted that formulas for the exact elastic analysis of an orthotropic deck can also be used for the analysis of cellular decks that are symmetric about the middle plane, i.e. the top and bottom plates have equal thicknesses. All that is required is to modify the input data for an orthotropic deck so as to produce a condition of anti-symmetry with respect to the middle plane as follows : 1) use only the antisymmetric component of the top and bottom stringer line loads (the symmetric component only squeezes the stringers and can be ignored) 2) use one half the actual flexural rigidity of the stringers B and B^b, and diaphragms, B^d ; and 3) use radius of gyration, ρ and ρ^b equal to zero (or if stringer representation is flexural rigidity and cross sectional area use an area approaching infinity).

The exact elastic model, Eqs. 238-241 for an orthotropic deck with flexible side supports can be formally reduced to a single equation with one unknown by successive elimination as was done with the two equation model for the simple side support case in Lecture VII ; however, the results for the four equation model are unwieldy and many of the coefficients are sums of obscure physical significance. (Even in the simpler case of simultaneous algebraic equations, it is seldom practical to derive an explicit formula solution for a model with more than three equations). For this model, convergence of the series summed on J is very rapid and indications are that computers, or programmable calculators will normally be used to get numerical results ; thus, an alternate procedure is recommended as follows :

1) Truncate the series on H_{kj} and $\overset{\$}{W}_j$ in Eq. 239 after one term (i.e. use J = 0 only) and solve eqs. 238 and 239 simultaneously for $Y_{k\ell}$ and $R_{k\ell}$ ($\ell < n$) in terms of $N_{k\ell}$, $\overset{\$}{V}_\ell$ (or $V_\ell^{a/s}$) and $\overset{\$}{W}_\ell$ (or $W_\ell^{a/s}$).
2) Substitute the results of step 1 into Eqs. 240 and 241 and solve the resulting algebraic equation for $\overset{\$}{V}_\ell$ (or $V_\ell^{a/s}$) and $\overset{\$}{W}_\ell$ (or $W_\ell^{a/s}$).
3) Substitute results of step 2 into the results of step 1 to find $Y_{k\ell}$ and $R_{k\ell}$.
4) Use the cyclic properties of R_{kj} (e.g. $R_{k\ell} = R_{k,2n+\ell} = R_{k,\ell-2n}$) to solve Eqs. 238 and 240 for the higher harmonics of Y_{kj} and $\overset{\$}{V}_j (j > n)$ and then substitute into Eq. 241 to find the higher harmonics of $\overset{\$}{W}_j$. (That is, first use Eq. 238 to eliminate Y_{kj} from Eq. 240 and solve for $\overset{\$}{V}_j$ (or $V_j^{a/s}$). Then find Y_{kj} from Eq. 238 and, in turn, $\overset{\$}{W}_j$ from Eq. 241).
5) If unusual accuracy is required, retain additional terms in the summation of Y_{kj} and $\overset{\$}{W}_j$ in Eq. 239 (J = − 2 to + 2 is sufficient), solve for improved results for $R_{k\ell}$ and repeat step 4. (In most cases, the results obtained in step 4 on the initial cycle are sufficiently accurate so that step 5 can be omitted).

This completes the algorithm for the general case of flexible side supports. Note that the effects of the side boundary deflections invalidate the one-to-one relation between the "k" loading and solution harmonics that existed for the case of simple side supports. For example, a single "k odd" loading harmonic will normally cause a deflection field with series coefficients containing all possible k odd harmonics. The relation between the "j" (second index) loading and solution harmonics is as described in the section on simple side supports.

In order to illustrate numerical use of the general bridge deck formulas under loading conditions which place a severe test on the convergence of the solution series, consider a bridge with the following physical parameters and loading: a = 360 in., b = 720 in.; m = 12; n = 4; t = .375in.; μ = .29; E = 29000 ksi; $B = B^{b}$ = 10.2681×10^{6} kip in.^{2}; $e = e^{b}$ = 14.4286 in., $\rho = \rho^{b}$ = 5.807 in. and B^{d} = 8.41×10^{6} kip in.^{2}. The loading consists of two symmetrically placed 20 kip concentrated loads, i.e. $N(r, y) = 20(\delta_r^5 + \delta_r^7)\,\delta(y - b/2)$ or $N_{kj} = 8P/mb\,(-1)^{\frac{k-1}{2}}(-1)^{\frac{j-1}{2}} \cos k\pi/m$ (k and j odd only) and $P_j^{\$} = P_j^{a\$} = 0$.

The combination of a relatively small number of stringers and diaphragms and loads of infinite intensity tend to show a harmonic analysis in a poor light due to slow convergence; however, as the results below indicate, even for this case the convergence is quite good, yielding practical results after only a small number of terms.

Some of the intermediate results are:

K = 11,873.kip/in.(Eq.A-1); $\overset{*}{A}_1$ = 268.68 in^2/kip, $\overset{*}{B}_1$ = 1.2374 in.^{2} kip and $\overset{*}{D}_1$ = 16.915 in.^{2}/kip (Eq.A-27); B_{11} = 0.07295 in.^{2}/kip (Eq. A-23); W_{11}^{N} = 4.8088 in.(Eq.15); A_1 = 1.494 in./kip, (Eq.A-29). A_1^{d} = .6834 in./kip.,(Eq.19); $\bar{B}_{11}$ = 2.2456 (Eq.A-24); and $\bar{T}_1^{\$}$ = - 26.179 ksi (Eq.A-15), $V_1^{\$}$ = .01530 in., $W_1^{\$}$ = .2818 in. (step 2 of algorithm p.108); H_{11} = .1705 kip/in., R_{11} = .7203 kip (step 3 of algorithm p. 108), $V_7^{\$}$ = - 1.123×10^{-5} in., H_{17} = -.02362 kip/in.; $W_7^{\$}$ = - 5.213×10^{-5} in., (step 4 of algorithm p.108); and R_{11} = .7195 kip (step 5 of algorithm p. 108 which confirms that recycling is unnecessary).

The deflection fields are as follows:

$$v(r,y) = \left[23.97 \cos\frac{\pi y}{b} - 1.318 \cos\frac{3\pi y}{b} + .2976 \cos\frac{5\pi y}{b} + \ldots\right] 10^{-3} \sin\frac{\pi r}{m}$$

$$+ \left[5.774 \cos\frac{\pi y}{b} + .7076 \cos\frac{3\pi y}{b} - .1253 \cos\frac{5\pi y}{b} + \ldots\right] 10^{-3} \sin\frac{3\pi r}{m}$$

$$+ \left[3.261 \cos\frac{\pi y}{b} + .01996 \cos\frac{3\pi y}{b} - .00567 \cos\frac{5\pi y}{b} + \ldots\right] 10^{-3} \sin\frac{5\pi r}{m} + \ldots$$

$$w(r,y)=\left[848.0\sin\frac{\pi y}{b}-14.20\sin\frac{3\pi y}{b}+1.876\sin\frac{5\pi y}{b}+\ldots\right]10^{-3}\sin\frac{\pi r}{m}$$

$$+\left[92.89\sin\frac{\pi y}{b}+.7133\sin\frac{3\pi y}{b}-.8491\sin\frac{5\pi y}{b}+\ldots\right]10^{-3}\sin\frac{3\pi r}{m}$$

$$+\left[62.07\sin\frac{\pi y}{b}-.1742\sin\frac{3\pi y}{b}+.01886\sin\frac{5\pi y}{b}+\ldots\right]10^{-3}\sin\frac{5\pi y}{m}+..$$

The membrane stress resultant field, n_y .(from Eq. A-1) is

$$n_y(x,y)=-K\left\{\left[.0944\sin\frac{\pi y}{b}-.01433\sin\frac{3\pi y}{b}+.005499\sin\frac{5\pi y}{b}+\ldots\right]10^{-3}\sin\frac{\pi x}{a}\right.$$

$$+\left[.02460\sin\frac{\pi y}{b}+.003893\sin\frac{3\pi y}{b}-.002209\sin\frac{5\pi y}{b}+\ldots\right]10^{-3}\sin\frac{3\pi x}{a}$$

$$+\left[.01533\sin\frac{\pi y}{b}+.00067\sin\frac{3\pi y}{b}-.000239\sin\frac{5\pi y}{b}+\ldots\right]10^{-3}\sin\frac{5\pi x}{a}+\ldots$$

The membrane stress resultant, n_y , at the center of the deck ($x = a/2$, $y = b/2$) is $n_y = 1.293$ kips/in. The finite element analysis described in the following paragraphs yields a stress $n_y = 1.179$, 1.192 or 1.419 kips/in. depending upon type of element utilized.

For comparative purposes the bridge system analyzed for Example 4 was also analyzed by use of a more comprehensive theoretical model and by use of discretized or finite element model.

The more comprehensive theoretical model was one which included the out-of-plane stiffness or flexural actions of the deck plate as well as its in-plane stiffness. The composite membrane-flexural model treated N(r ,y) as an unknown out-of-plane interactive force between the stringers and the plate and rationally accounted for the effects of deck loads applied between stringers. The computations were thus complicated considerably but gave essentially the same deflection field, for example the maximum difference for w(r, y) was 2.3 % which confirmed the authors'hypothesis that the composite membrane model (238-241) is sufficiently sophisticated to analyze metal deck bridges of orthotropic design.

There was also some question as to the need for a rational theoretical analysis in view of the availability of various open form finite element programs which can be modified to approximately model such decks. A space frame program (for the stringers, diaphragms and pseudo stud members of length e to model composite action) was combined with a finite element plane stress program (using elements) whose width equaled the stringer spacing and length equaled 1/3 the diaphragm spacing) to analyze the deck as an "equivalent" framework. Even though double symmetry was utilized, this relatively coarse network required two orders of magnitude more computing time than did the formula approach (which incidentally was written to give research accuracy rather than computational efficiency) and, of more significance, required nearly three orders of magnitude more input information (only one card is needed to read in data for the theoretical approach). The finite element results were in error by up to 10% for deflections and the plate stress distributions bore little resemblance to the exact results. The need for rationally based formulas appeared to be confirmed.

The Waffle Plate with Flexible Side Supports (*). – As a fifth and final structure with "difficult" boundary conditions, we will consider a system that is more deeply into the category of two-dimensional lattices with two-dimensional elements than was the case with the preceding example. The waffle plate model denotes a continuous plate in continuous contact with (and usually monolithic with) two sets of reinforcing beams. For the purpose of this analysis, we will consider a somewhat specialized version of the waffle plate such that the only interior interactive forces between the continuous plate and the reinforcing ribs are the discrete-continuous out-of-plane forces between the stringers and the plate N(r, y) and between the diaphragms and the plate M(x, s). These mixed discrete-continuous out-of-plane interactive forces are in contrast to the in-plane discrete-continuous forces Y(r, y) and the out-of-plane discrete forces R(r, s) in the preceding plate-stringer-diaphragm problem. The specialized waffle plate with only out-of-plane rib-plate interactive forces is an accurate model for two-way reinforced plates with various type of detailing and/or member property ratios. One system which can be so modeled is the typical concrete bridge deck resting on and in continuous contact with a stringer-diaphragm support grid in which no provision was made for shear lugs to obtain composite action. A second system conforming to such a model is a monolithic grid-plate structure in which the middle plane of the plate also includes

(*) The analysis of this structure was first published in the following reference: Dean, D.L. and Avent, R.R., "Analysis of Metal Plate-Stringer Diaphragm Bridge Decks," *International Association for Bridge and Structural Engineering,* Vol. 35-I, April 1975, pp. 45-64.

the centroidal axes of the grid members (see Fig. 19) and the member properties are

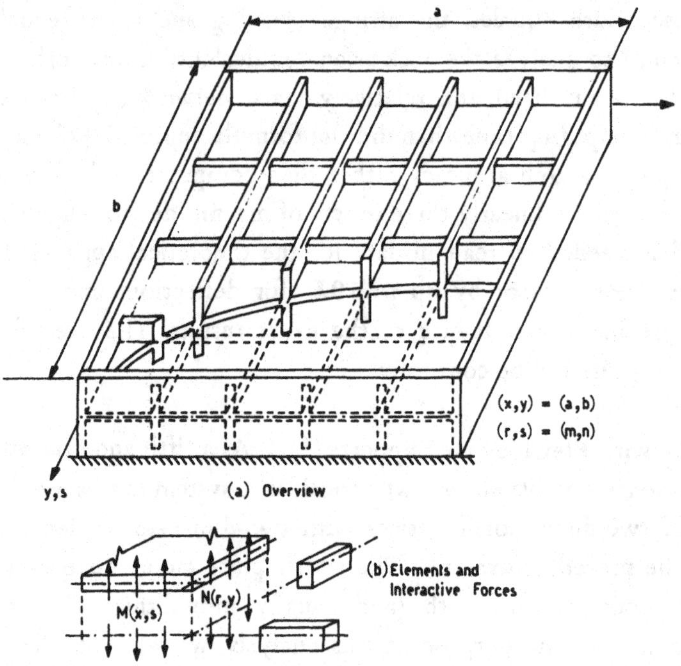

Fig. 19. Flexural Waffle Plate

such that the effects of torsional interactive forces are negligible. (Note, that the question of composite action does not arise due to zero eccentricity of any in-plane plate forces).

The plate formulas required for analysis of this special type of waffle plate are shown in Appendix V, i.e., the behavior is that of a flexural plate rather than a membrane. The continuous plate deflection field is found from :

$$w(x,y) = w^h(x,y) + w^a(x,y) - \sum_{\beta=1}^{n-1} \int_0^a M(\xi,\beta) K^{wz}\left(x,y,\xi,\frac{b}{n}\beta\right) d\xi$$

$$- \sum_{\alpha=1}^{m-1} \int_0^b N(\alpha,\eta) K^{wz}\left(x,y,\frac{a}{m}\alpha,\eta\right) d\eta \qquad (242)$$

in which : w^h is the homogeneous solution due to the side boundary deflections (see Eqs. A-5.14 thru A-5.17), w^a is the particular solution due to the applied loading on a simply supported plate (see Eqs. A-5.6 thru A-5.10) ; and $M(\xi,\beta)$ and $N(\alpha,\eta)$ are the unknown plate-rib interactive forces defined above. The plate deflection series is :

$$w(x,y) = w^h(x,y) + \sum_{i=1}^{\infty}\sum_{j=1}^{\infty}\left[\overset{*}{q}_{ij} - \frac{m}{a}N_{ij} - \frac{n}{b}M_{ij}\right]\overset{*}{C}_{ij} \sin\alpha_i x \sin\bar{\alpha}_j y \tag{243a}$$

$$M(x,s) = \sum_{i=1}^{\infty}\sum_{\ell=1}^{n-1} M_{i\ell} \sin\frac{\ell\pi s}{n} \sin\alpha_i x \tag{243b}$$

$$M_{i\ell} = M_{i,2Jn+\ell} = -M_{i,2Jn-\ell} \tag{243c}$$

$$N(r,y) = \sum_{j=1}^{\infty}\sum_{k=1}^{m-1} N_{kj} \sin\frac{k\pi r}{m} \sin\bar{\alpha}_j y \tag{243d}$$

$$N_{kj} = N_{2Im+k,j} = -N_{2Im-k,j} \tag{243e}$$

$w^h(x,y)$ is given by Eq. A-5.15 or A-5.17 ; and $\overset{*}{q}_{ij}$ and $\overset{*}{C}_{ij}$ are given by Eqs. A-5.10 and A-5.7 respectively.

The unknowns are the coefficients of the interactive forces $M_{i\ell}$ and N_{kj} and the coefficients of the boundary displacements W_j^s and/or $W_j^{a/s}$ (see Eqs. A-5.14 or Eq. A-5.16). The compatibility equations required to find these coefficients are found by matching plate to stringer displacements along $0 < y < b$ for $r = 1, (1), m-1$, plate to diaphragm displacements along $0 < x < a$ for $s = 1, (1), n-1$ and by matching plate edge displacements along $x = 0$ and a to the boundary beams, which may also be labeled by the discrete coordinates as $r = 0$ and m. The required plate stringer line and stringer displacements parallel to the y axis are :

$$w\left(\frac{a}{m}r, y\right) = \sum_{j=1}^{\infty}\sum_{k=1}^{m-1}\left[W_{kj}^{a} + \frac{4}{m}\bar{W}_{kj}\bar{C}_{kj} - N_{kj}C_{kj} - W_{kj}^{M}\right]\sin\frac{k\pi r}{m}\sin\bar{\alpha}_j y \tag{244a}$$

$$W_{kj}^{M} = \frac{n}{b}\sum_{I=-\infty}^{+\infty} M_{2Im+k,j}\overset{*}{C}_{2Im+k,j} \tag{244b}$$

in which W_{kj}^{a} , $\bar{W}_{kj}$, $\bar{C}_{kj}$, and C_{kj} are given by Eqs. A-5.29, A.5-16, A.5-27 and A.5-24 respectively.

$$w_s(r,y) = \sum_{k=1}^{m-1}\sum_{j=1}^{\infty} \overset{*}{A}_j N_{kj} \sin\frac{k\pi r}{m} \sin\bar{\alpha}_j y \tag{245}$$

in which $\overset{*}{A}_j$ is given by Eq. A-4.23.

Thus one of the compatibility equations is obtained by matching coefficients in Eqs. 244 and 245, i.e.

$$(\overset{*}{A}_j + C_{kj})N_{kj} + \frac{n}{b}\sum_{I=-\infty}^{+\infty} M_{2Im+k,j}\,\overset{*}{C}_{2Im+k,j} - \frac{4}{m}\bar{C}_{kj}\bar{W}_{kj} = W^a_{kj} \tag{246}$$

Similarly, the plate diaphragm line and diaphragm displacements parallel to the x axis are :

$$w\left(x,\frac{b}{n}s\right) = \sum_{i=1}^{\infty}\sum_{\ell=1}^{n-1}\left[\underline{W}^a_{i\ell} + \underline{W}^w_{i\ell} - W^N_{i\ell} - M_{i\ell}\,\underline{C}_{i\ell}\right] \sin\alpha_i x \sin\frac{\ell\pi s}{n} \tag{247a}$$

in which the coefficients are defined similarly to those in Eq. 244, i.e.

$$\underline{W}^a_{i\ell} = \sum_{J=-\infty}^{+\infty} \overset{*}{q}_{i,2Jn+\ell}\,\overset{*}{C}_{i,2Jn+\ell} \tag{247b}$$

$$\underline{W}^w_{i\ell} = \frac{4}{a}\sum_{J=-\infty}^{+\infty} \overset{*}{\bar{C}}_{i,2Jn+\ell}\,\bar{W}_{i,2Jn+\ell} \tag{247c}$$

$$W^N_{i\ell} = \frac{m}{a}\sum_{J=-\infty}^{+\infty} \overset{*}{C}_{i,2Jn+\ell}\,N_{i,2Jn+\ell} \tag{247d}$$

$$\underline{C}_{i\ell} = \frac{n}{b}\sum_{J=-\infty}^{+\infty} \overset{*}{C}_{i,2Jn+\ell} = \frac{b}{4nD\alpha_i^2 D_{i\ell}}\left[\frac{\sinh\underline{\lambda}_i}{\underline{\lambda}_i} - \frac{1-\cosh\underline{\lambda}_i\cos\frac{\ell\pi}{n}}{D_{i\ell}}\right] \tag{247e,f}$$

(247g,h) $\underline{\lambda}_i = \frac{b}{n}\alpha_i$, $D_{i\ell} = \cosh\underline{\lambda}_i - \cos\frac{\ell\pi}{n}$ and

$$w_D(x,s) = \sum_{i=1}^{\infty}\sum_{\ell=1}^{n-1}\left(\overset{*}{\underline{A}}_i M_{i\ell} + \frac{4}{a}\frac{1}{\alpha_i}\underline{W}_{i\ell}\right) \sin\alpha_i x \sin\frac{\ell\pi s}{n} \tag{248a}$$

in which the coefficients are defined similarly to those previously introduced, i.e.

$$\overset{*}{\underset{\sim i}{A}} = \frac{1}{B^d \alpha_i^4}, \quad B^d = \text{flexural rigidity of diaphragms} \tag{249b}$$

$$\underset{\sim i\ell}{W} = \sum_{J=-\infty}^{+\infty} \bar{W}_{i,2Jn+\ell} = \text{coefficients of boundary node deflections} \tag{249c}$$

$$\frac{4}{a} \sum_{i}^{\infty} \frac{1}{\alpha_i} \sin \alpha_i x = \begin{bmatrix} 1 & \text{for } i \text{ odd} \\ 1 - \frac{2x}{a} & \text{for } i \text{ even} \end{bmatrix} \tag{249d}$$

Thus the second compatibility equation is obtained by matching coefficients in Eqs. 247 and 248, i.e.

$$(\overset{*}{\underset{\sim i}{A}} + \underset{\sim i\ell}{C})M_{i\ell} + \frac{m}{a} \sum_{J=-\infty}^{+\infty} \overset{*}{C}_{i,2Jn+\ell} N_{i,2Jn+\ell} - \frac{4}{a} \sum_{J=-\infty}^{+\infty} \left(\overset{*}{\bar{C}}_{i,2Jn+\ell} - \frac{1}{\alpha_i} \right) \bar{W}_{i,2Jn+\ell} = \underset{\sim i\ell}{W^a} \tag{250}$$

The third equation required to solve for the unknown series coefficients is obtained by expressing the boundary beam deflection in terms of the loads applied to it including the plate edge shear resultants. The loads applied directly to the boundary beams consist of the given loads :

$$P\left(\begin{smallmatrix}0\\m\end{smallmatrix}, y\right) = \sum_{j=1}^{\infty} (P_j^{\$} \pm P_j^{a/s}) \sin \bar{\alpha}_j y \tag{251a}$$

and the set of concentrated reactions from the diaphragms

$$R\left(\begin{smallmatrix}0\\m\end{smallmatrix}, s\right) = \sum_{s=1}^{n-1} (R_\ell^{\$} \pm R_\ell^{a/s}) \sin \frac{\ell \pi s}{n} \tag{251b}$$

$$\sum_{i}^{\infty} \frac{1}{\alpha_i} M_{i\ell} = \begin{bmatrix} R_\ell^{\$} & \text{for } i \text{ odd} \\ R_\ell^{a/s} & \text{for } i \text{ even} \end{bmatrix} \tag{251c}$$

The plate edge shear resultants consist of those components due to :

a) the applied distributed load, q(x, y), on the simply supported plate ;

b) the plate-stringer interactive forces N(r, y) ;
c) the plate-diaphragm interactive forces M(x, s) ; and the side boundary displacements $w^h(\overset{0}{a},y)$. Matching coefficients of the infinite series for the various symmetric and anti-symmetric components of the boundary beam displacements yield the following :

$$\frac{1}{\overset{*}{A}{}^b_j} W^\$_j = P^\$_j + \frac{n}{b} R^\$_j + S^{a\$}_j - S^{N\$}_j - S^{M\$}_j + S^{h\$}_j W^\$_j \tag{252a}$$

$$\frac{1}{\overset{*}{A}{}^b_j} W^{a/s}_j = P^{a/s}_j + \frac{n}{b} R^{a/s}_j + S^{aa/s}_j - S^{Na/s}_j - S^{Ma/s}_j + S^{ha/s}_j W^{a/s}_j \tag{252b}$$

in which $S^{a\$}_j$ and $S^{aa/s}_j$ and $S^{h\$}_j$ and $S^{ha/s}_j$ are given by Eqs. A-5.32 and A-5.21 respectively and the coefficients due to the interactive forces are

$$\begin{bmatrix} S^{N\$}_j \\ \\ S^{Na/s}_j \end{bmatrix} = \frac{m}{a} \sum_i^{\infty} \overset{*}{\bar{C}}_{ij} N_{ij} = \sum_k^{m-1} \bar{C}_{kj} N_{kj} \qquad \begin{bmatrix} \text{for i,k odd} \\ \\ \text{for i,k even} \end{bmatrix} \tag{253a,b}$$

in which $\overset{*}{\bar{C}}_{ii}$ and $\bar{C}_{kj}$ are given by Eqs. A-5.13 and A-5.27 respectively.

$$\begin{bmatrix} S^{M\$} \\ \\ S^{Ma/s} \end{bmatrix} = \frac{n}{b} \sum_i^{\infty} \overset{*}{\bar{C}}_{ij} M_{ij} \qquad \begin{bmatrix} \text{for i odd} \\ \\ \text{for i even} \end{bmatrix} \tag{254}$$

Thus the third and final equation (shown here for symmetric components only,Eq. 255b denotes the corresponding equation for anti-symmetric components) is :

$$\left[(\overset{*}{A}{}^b_j)^{-1} - S^{h\$}_j\right] W^\$_j + \sum_{k=1,3...}^{m-1} \bar{C}_{kj} N_{kj} + \frac{n}{b} \sum_{i=1,3...}^{\infty} \left(\overset{*}{\bar{C}}_{ij} - \frac{1}{\alpha_i}\right) M_{ij} = P^\$_j + S^{a\$}_j \tag{255a}$$

Thus, it is required to solve Eqs. 255, 250 and 246 simultaneously for the unknown coefficients N_{kj} , $M_{i\ell}$ and $W^\$_j / W^{a/s}_j$.

As in the previous problems in this lecture, the simultaneous equations are not conformable, i.e., each term in Eq. 246 is indexed kj, each term in Eq. 250 is indexed $i\ell$ and each term in Eq. 255 is indexed j. Also each equation contains at least one infinite sum on one of the unknown coefficients. Thus, we will describe a rapidly convergent algorithm for solving the three simultaneous equations rather than give closed formulas for the coefficients, as the latter is not yet available for this class of simultaneous summation equations model. The recommended algorithm is as follows :

1. Find approximate results for the first m and n harmonics of $M_{i\ell}$, N_{kj} and $W_j^\$$ by assuming the higher harmonics to be zero. Specifically :
   a) use I & J = 0 only in Eqs. 246 and 250 which are then solved for $M_{i\ell}$ and N_{kj} in terms of the applied loads and $W_j^\$$ for i, k = 1, (2), m – 1 and j, ℓ = 1, (2), n – 1.
   b) Substitute the results of 1.a) into Eq. 255 (truncating sum on i at m – 1) and solve for the first cycle values of $w_j^\$$ for j = 1, (2), n – 1
   c) Substitute the results of 1.b) back into the results of 1.a) to find first cycle values for the lower harmonics of $M_{i\ell}$ and N_{kj}.
2. Use the results of step 1 to find approximate values for all of the desired higher harmonics of $M_{i\ell}$, N_{kj} and $W_j^\$$. Specifically :
   a) Still using J = 0 only, substitute the results for $W_j^\$$ and N_{kj} from step 1 into Eq. 250 and solve for the desired higher i harmonics of $M_{i\ell}$.
   b) Substitute the results of 2a) into Eq. 246 to solve for N_{kj} in terms of $W_j^\$$ for all desired harmonics.
   c) Substitute the results of 2.a) and 2.b) into Eq. 255 and solve for all desired harmonics of $W_j^\$$.
   d) Substitute the results of 2.c) back into the results of 2.b) to find corresponding values for N_{kj} .
3.) Substitute the results of steps 1 and 2, back into the appropriate equations to find improved values for $M_{i\ell}$, N_{kj} and $W_j^\$$ and repeat until desired accuracy is obtained : Specifically :
   a) Substitute the values for N_{kj} and $W_j^\$$ into Eq. 250 to find the next cycle values for $M_{i\ell}$.
   b) Use these values of $M_{i\ell}$ and above values of $W_j^\$$ in Eq. 246 to find the next cycle values for N_{kj} .
   c) Use the results of 3.a) and 3.b) in Eq. 255 to find the next cycle values for $W_j^\$$.

This completes the algorithm for solving the model for the flexural waffle plate with flexible side supports. For many cases, the deflection fields are satisfactorily determined by using only step 1.

It should be noted that the interior unknowns for this model of the waffle plate consist only of the two fields of mixed discrete-continuous rib-plate interactive forces. In contrast, the interior unknowns for the plate-stringer-diaphragm system consisted of a mixed discrete-continuous stringer-plate interactive force field and a pure discrete stringer-diaphragm interactive force field whose coefficients were denoted $R_{k\ell}$. One might ask why there is no $R_{k\ell}$ here due to the interaction of the two sets of ribs (*). If the approach were to first analyze the two sets of ribs as a grid (which requires the determination of $R_{k\ell}$), and then to establish compatibility between the grid and the plate, the discrete interactive forces would be introduced. For the above model however, the concept was to establish compatibility between the plate and the two sets of ribs independently ; this did, however, implicity establish node deflection compatibility between the two sets of ribs which would be the requirement for finding $R_{k\ell}$ if R(r, s) were established as a separate unknown. In other words, the total forces "felt" by one set of ribs – say the stringers, N(r, y), would be the same in both cases, but with the grid-plate approach, the set of concentrated forces R(r, s) would be removed from N(r, y) and handled separately, probably resulting in a smoother (or more rapidly convergent) expression for the remaining mixed discrete-continuous forces N'(r, y). That is,

$$N(r,y) = N'(r,y) - \sum_{\beta=1}^{n-1} R(r,\beta)\,\delta\left(y - \frac{b}{n}\beta\right) \text{ or } N_{kj} = N'_{kj} - \frac{n}{b} R_{kj} .$$

In order to illustrate numerical use of this model and algorithm, consider the noncomposite flexural waffle plate with the following physical parameters and loading : $a = 2/3b = 480$ in., $m = 8$, $n = 4$, $B = bD = 7.2 \times 10^9$ lb.in^2; $B^d = .2B$; $B^b = 3B$ $\mu = 0.2$ $\overset{*}{q}_{ij} = q_o \delta_i^1 \delta_j^1$, $q_o = 1.0$ psi (equivalent to single term expansion for uniform load equal to 88.8 psf or 4.25 kn/m^2), $P_j^{\$} = P_j^{a/s} = 0$. For such a harmonic loading the solution coefficients indexed ℓ are zero except for $\ell = 1$ and those indexed j are zero except for $j = 2Jn \pm 1$. Also the $i = k = 1$ and $j = \ell = 1$ terms are dominant in step 1 of the algorithm which was modified by treating the series in Eq. 255 as one term so that Eqs. 246,

---

(*) **The analysis of this structure, and a broad class of related reinforced plate problems, was first published in the following reference: Payne, William W., Jr., "Analysis of Rectangular Plates with Two Sets of Composite Ribs", thesis submitted for Ph. D. Degree in Civil Engineering, North Carolina State University, Raleigh, N.C., 1976.**

250 and 255 could be solved as simultaneous algebraic equations for the first cycle approximation to N_{11}, M_{11}, and $\overset{\$}{W_1}$. Some of the intermediate results were : $\overset{*}{C}_{1,1} = 9.7193 \times 10^{-4}\ \frac{b^4}{D}$; $\overset{a}{W}_{kj} = \overset{a}{\underline{W}}_{i\ell} = \overset{*}{C}_{1,1}\, q_o \delta^1_{k,i} \delta^1_{\ell,j}$; $C_{1,1} + \overset{*}{A}_1 = 2.1930 \times 10^{-2}\ \frac{b^3}{D}$ $\underline{C}_{1,1} + \overset{*}{\underline{A}}_1 = 1.4050 \times 10^{-2} \frac{b^3}{D}$; $\overset{*}{\bar{C}}_{1,1} - \frac{1}{\alpha_1} = -\ 2.9131 \times 10^{-2}\, b$ $\bar{C}_{1,1} = 2.1641$; $\overset{a\$}{S_j} = \overset{*}{\bar{C}}_{1,1}\, q_o \delta^1_j$; $\overset{*}{\bar{C}}_{1,1}\, q_o = .183075\ q_o b$ $\overset{h\$}{S_1} = -26.9278\ \frac{D}{b^3}$; and $\overset{*b}{A_1} = 3.4220 \times 10^{-3} b^3/D$. Substitution of these coefficients and solution for the first approximation for N_{11}, M_{11} and $\overset{\$}{W_1}$ (using $b = 720$ in. $D = 10^7$ lb.in and $q_o = 1$ psi.) give the following results.

$N_{1,1} = 5.1620 \times 10^{-2}\, q_o b = 37.166$ lb./in; $M_{1,1} = 2.3437 \times 10^{-2}\, q_o b = 16.875$ lb./in. and $\overset{\$}{W_1} = 2.3216 \times 10^{-4}\ \frac{q_o b}{D} = 6.239$ in.

These results were substituted back into the original equation with untruncated summations and cycled to find the higher harmonics as described in Steps 2 and 3 with the following results :

$\overset{\$}{W_1} = 5.8245$ in.; $\overset{\$}{W_7} = -\ 5.914 \times 10^{-4}$; $\overset{\$}{W_9} = 2.1411 \times 10^{-4}$; $\overset{\$}{W}_{15} = -2.6834 \times 10^{-5}$ and $\overset{\$}{W}_{17} = 1.610 \times 10^{-5}$; $N_{11} = 36.491\ \frac{\text{lb}}{\text{in.}}$; $N_{31} = 5.558$; $N_{51} = 2.525$; $N_{71} = .7535$; $N_{17} = 5.225$; $N_{37} = -\ 2.958$; $N_{57} = -\ 1.750$; $N_{77} = -\ .597$; $N_{19} = -5.085$ $N_{39} = 2.964$; $N_{59} = 1.802$, $N_{79} = .637$, $N_{1,15} = 4.087$, $N_{3,15} = -2.782$; $N_{5,15} = -1.626$ $N_{7,15} = -\ .554$; $N_{1,17} = -3.645$; $N_{3,17} = 2.700$; $N_{5,17} = 1.549$; $N_{7,17} = .518$, etc. $M_{11} = 17.599\ \frac{\text{lb}}{\text{in.}}$, $M_{31} = -\ 10.321$; $M_{51} = -7.691$; $M_{71} = -5.640$; $M_{91} = -3.232$; $M_{11,1} = -\ 4.210$; $M_{13,1} = 3.309$; $M_{15,1} = 8.877$; $M_{17,1} = -14.032$; $M_{19,1} = -9.319$; $M_{21,1} = -\ 5.283$; $M_{23,1} = -\ 1.984$; $M_{25,1} = 1.188$, etc.

These final results can be substituted into the various expressions for deflection fields to find all the other desired system descriptors. Of course the deflection series converge more rapidly than the series for interactive forces. This completes the example for the noncomposite flexural analysis of a waffle plate with flexible side supports.

APPENDIX I

ORDINARY DIFFERENCE EQUATIONS

A kth order ordinary difference equation with constant coefficients can be written in recurrence form as follows:

$$f(r+k) + A_{k-1} f(r+k-1) + \ldots A_2 f(r+2) + A_1 f(r+1) + A_0 f(r) = V(r)$$

(A-1.1a)

The equation can be most easily rewritten in operator form by use of the displacement operator(*) i.e.,

(A-1.1b) $(E_r^k + A_{k-1} E_r^{k-1} + \ldots A_2 E_r^2 + A_1 E_r + A_0) f(r) = V(r)$

Homogeneous Solutions – The homogeneous part of Eq. A-1.1b can be written as

(A-1.2a) $$\psi(E_r) f^h(r) = 0$$

from which it may be seen substitution of $f^h(r) = \beta^r$ gives

(A-1.2b) $$\psi(\beta)\beta^r = 0$$

or

(A-1.3) $$f^h(r) = C_1\beta_1^r + C_2\beta_2^r + \ldots C_k\beta_k^r$$

in which β_1, β_2 ... β_k are simple roots(**) of the polynomial ψ and the

---

(*) The analyst first encountering difference equations after a typical background in differential equations where only the differential operator is used, tends to become bewildered by the variety of operators used in the calculus of finite differences. Any of the first order operators shown in Table A-1 in Appendix II (except the central difference and central mean operators δ and μ which are used primarily for the approximate numerical analysis of continuum models by finite differences) could be used exclusively; however, the concept of the differences Δ and ∇ and the means **N** and **И** are such that the experienced analyst tends to use the different symbols rather than their equivalents in terms of say the displacement operator E. The second central difference and the second mean-difference operators occur so frequently in structural models that use of their special symbols ◸ and ⊟ also become attractive.

(**) The solution for a double root, say for $\beta_4 = \beta_5$, would be $(C_4 + C_5 r)\beta_4^r$ and higher degree coefficients would serve roots of higher multiplicity.

constants $C_1, C_2 \ldots C_k$ are determined so that the total solution satisfies the boundary conditions. While the solution form in Eq. A-1.3 is of fundamental importance, it is not often used in practice due to the lack of symmetry of exponential functions and the necessity of using complex variable theory for complex roots. In the great bulk of physical problems the operator polynomial is symmetric; i.e., $A_0 = 1$, $A_{k-1} = A_1$, etc. so that the equation can be written in terms of even differences; i.e.

$$(\nabla_r^{k'} + A'_{k'-1}\nabla_r^{k'-1} + \ldots A'_1\nabla_r + A'_0) f^h(r) = 0 \qquad \text{(A-1.4a)}$$

$$k' = \frac{1}{2} k$$

or

$$(\nabla_r + 2\gamma_1)(\nabla_r + 2\gamma_2) \ldots (\nabla_r + 2\gamma_{k'}) f^h(r) = 0 \qquad \text{(A-1.4b)}$$

Thus, for various ranges of real γ's the solution can be written as a sum of the solution types shown at the top of Table A-2 in Appendix II. To complete the coverage of symmetric ordinary difference equations, the case where two of the roots in Eq. A-1.4b appear as a pair of complex conjugates must be considered; i.e.,

$$(\nabla_r^2 + 4g\nabla + 4h) F(r) = 0 \qquad \text{(A-1.5a)}$$

or

$$(\nabla_r + 2\gamma)(\nabla_r + 2\bar{\gamma}) F(r) = 0 \qquad \text{(A-1.5b)}$$

in which

$$\gamma, \bar{\gamma} = g \pm \sqrt{g^2 - h} \qquad \text{(A-1.5c)}$$

The possible cases of solutions for this symmetric quartic, for different ranges of the parameters, are:

1. $h \leqslant g^2$ in which case both γ and $\bar{\gamma}$ are real and the solution is simply a combination of the solutions to the symmetric quadratic shown in Table A-2 in Appendix II.
2. $h > g^2$ in which case the quadratic parameters are γ, $\bar{\gamma} = g \pm i\sqrt{h - g^2}$ and the solution is $F^h(r) = C_1 \cosh \eta r \cos \phi r$ (A-1.6)

$$+ C_2 \cosh \eta r \sin\phi r + C_3 \sinh \eta r \cos\phi r + C_4 \sinh \eta r \sin\phi r$$

a) for $g = 1$, $\phi = \pi/2$ and $\eta = \ln(\sqrt{h} - \sqrt{h-1})$

b) for $g \neq 1$

(A-1.7a) $$\eta = \ln \left[\frac{\zeta(g,h) + 1 - g}{1 - g - \zeta(g,h)} \right]^{\frac{1}{2}}$$

(A-1.7b) $$\phi = \cos^{-1} \left[\frac{1 - g}{\cosh \eta} \right]$$

(A-1.7c) $$\zeta(g,h) = \left[\frac{1}{2} \sqrt{h(4 - 4g + h)} + g(g-1) - \frac{1}{2} h \right]^{\frac{1}{2}}$$

This completes the listing of possible homogeneous solutions to constant coefficient ordinary difference equations. However, attention should be called to Table A-3 in Appendix II which contains a collection of useful homogeneous solutions to the equattion $(\Delta_r + 2\gamma)\, w(r) = 0$ for a variety of symmetric and anti-symmetric inhomogeneous boundary conditions. By noting the boundary values of the particular solution, one can select the coefficients of the homogeneous solution from such a table with a considerable savings in algebraic manipulations. Such tabulations are more helpful in discrete field derivations than for analogous problems in the continuum as multiple difference operations have much more of a tendency to generate algebraically complicated expressions than do multiple differentiations.

Particular Solutions – Techniques for finding particular solutions to difference equations are closely analogous to those used in differential equations; for example, a useful technique is the method of undetermined coefficients. If the form of the solution is known or suspected, it is substituted into the equation with unknown coefficients and the coefficients are determined so as to satisfy the equation. As an example, consider the case of an exponential loading term: i.e.

(A-1.8) $$\psi(E_r) f^P(r) = C_0 a^r$$

This case is quite useful for structural analysis. For example, $a = 1$ covers the case of a uniform loading while $a = -1$ covers the case of alternating loads often used to find maximum joint moments. Assuming a particular solution of the same form, i.e., $f^P(r) = A a^r$ gives the following formula:

$$f^p(r) = \frac{C_0 a^r}{\psi(a)} \qquad \psi(a) \neq 0 \tag{A-1.9a}$$

If a is a simple root of ψ, so that the assumed form is part of the homogeneous solution, then a limiting process is used to get

$$f^p(r) = \frac{C_0 r a^{r-1}}{D_a \psi(a)} \tag{A-1.9b}$$

If a is a double root of ψ,

$$f^p(r) = \frac{C_0 r(r-1) a^{r-2}}{D_a^2 \psi(a)} \qquad \text{etc.} \tag{A-1.9c}$$

Another useful technique for finding particular solutions is the symbolic method which requires use of inverse operators and partial fractions; i.e.,

$$f^p(r) = \frac{V(r)}{\psi(E)} = \left[\frac{b_1}{E_r - \beta_1} + \frac{b_2}{E_r - \beta_2} + \dots \frac{b_k}{E_r - \beta_k} \right] V(r) \tag{A-1.10a}$$

in which the b's are found by standard methods

$$b_i = \left. \frac{1}{D_a \psi(a)} \right|_{a = \beta_i} \tag{A-1.10b}$$

If V(r) is a finite algebraic polynomial carrying out the indicated division with the operator E replaced by $\Delta + 1$ leads to a direct solution. For transcendental functions the following formula(*) is often useful

$$\frac{1}{E_r - \beta} V(r) = \beta^{r-1} \Delta_r^{-1} [\beta^{-r} V(r)] \tag{A-1.11}$$

An important example of the use of Eq. A-1.11 is as follows:

$$(\nabla_r + 2\gamma) K^p(r,\alpha) = \delta_r^\alpha \tag{A-1.12a}$$

in which δ_r^α is the standard Kronecker delta used as an impulse load term in

(*) "Calculus of Finite Differences", Charles Jordan, Chelsea Publishing Company, New York, 1950, p. 565.

discrete field mechanics.

(A-1.12b) $$K^p(r,\alpha) = \frac{\sinh \lambda\ (r-\alpha)}{\sinh \lambda}\ \not{U}(r-\alpha)$$

(A-1.12c) $$\cosh \lambda = 1-\gamma \qquad \gamma < 0$$

(A-1.12d) $$\delta_r^{\alpha} = \begin{bmatrix} 0 & r \neq \alpha \\ 1 & r = \alpha \end{bmatrix}$$

(A-1.12e) $$\not{U}(r-\alpha) = \begin{bmatrix} 0 & r < \alpha \\ 1 & r \geqslant \alpha \end{bmatrix}$$

(A-1.12f) $$\nabla_r \not{U}(r-\alpha) = \delta_r^{\alpha}$$

The results for other ranges of γ may be obtained by modifying Eq. A-1.12b as follows:

2. For $\gamma > 2$ multiply solution by $-(-1)^{r-\alpha}$ and use $\cosh \lambda = \gamma - 1$;
3. For $0 < \gamma < 2$ replace the hyperbolic functions in the solution and definition of λ by the corresponding trigonometric functions;
4. For $\gamma = 0$, $K^p(r,\alpha) = (r-\alpha)\ \not{U}\ (r-\alpha)$; and
5. For $\gamma = 2$, same as 4) except multiplied by $-(-1)^{r-\alpha}$.

For equations whose operators contain repeated factors of the types shown in Eq. A-1.12a see one of the author's earlier papers(*)

The Summation or Inverse Delta Operation. – The inverse delta operation could be viewed as simply a solution to the difference equation $(E_r - 1)\ f(r) = V(r)$, and thus covered in previous sections; however, its significance in evaluating sums is such as to warrant special attention. As a preliminary to dealing with the inverse operation,

---

(*) "Analysis of Structural Nets", Donald L. Dean and Celina P. Ugarte, Publications International Association for Bridge and Structural Engineering, Vol. 23, December 1963, p. 89.

attention will be given to the delta operation and its relation to the differential operator in the continuum; i.e.,

$$D_x x^n = n\, x^{n-1} \qquad \text{(A-1.13)}$$

$$\Delta_r (r)_n = n(r)_{n-1} \qquad \text{(A-1.14a)}$$

in which the subscript n denotes a nth degree factorial; i.e.,

$$(r)_n = r(n-1)(r-2)\ldots(r-n+2)(r-n+1) \qquad \text{(A-1.14b)}$$

In other words, while algebraic polynomials are convenient for differential operations, the natural form for difference operations is that of a factorial polynomial. A brief listing of stirling numbers for making such transformations is given in Appendix II, Table A-4). Two examples of such transformations are as follows:

$$a\, r^2 + br + c = a(r)_2 + (a+b)(r)_1 + c$$

and

$$(r)_3 - 3r = r^3 - 3r^2 - r$$

Extensive use of such transformations was required to complete the listing of difference operations on algebraic expressions shown in Appendix II, Table A-5.

Use of symbolic methods are helpful for interpreting the inverse delta operation as a sum: i.e.,

$$\Delta_r^{-1} = \frac{1}{\Delta_r} = -\frac{1}{1-E_r} = -(1 + E_r + E_r^2 + \ldots) \qquad \text{(A-1.15a)}$$

thus

$$\Delta_r^{-1} f(r) = -\sum_{i=r}^{\infty} f(i) \qquad \text{(A-1.15b)}$$

or Δ_r^{-1} gives an indefinite sum

$$\Delta_r^{-1} f(r) = \sum f(r) + C \qquad \text{(A-1.16a)}$$

and the definite sum can be written

(A-1.16b) $$\Delta_r^{-1} f(r)\Big|_a^{b+1} = \sum_{r=a}^{b} f(r)$$

Two examples of the use of the inverse delta to determine definite sums are as follows:

$$1. \quad \sum_{x=1}^{n} x^2 - x = \Delta_x^{-1} [(x)_2]\Big|_1^{n+1}$$

(A-1.17)
$$= \left[\frac{1}{3} (x)_3\right]\Big|_1^{n+1}$$

$$= \frac{1}{3} (n+1)_3 = \frac{n}{3} (n^2 - 1)$$

$$2. \quad \sum_{r=1}^{n-1} \sin \frac{k\pi r}{n} = \Delta_r^{-1} \sin \frac{k\pi r}{n}\Big|_1^{n}$$

or from Table A-2 in Appendix II

(A-1.18)
$$= -\frac{1}{2 \sin \frac{k\pi}{2n}} \cos \frac{k\pi}{n}\left(r - \frac{1}{2}\right)\Big|_1^{n}$$

$$= \begin{bmatrix} \cot \frac{k\pi}{2n} & k \text{ odd} \\ 0 & k \text{ even} \end{bmatrix}$$

APPENDIX II

TABLES OF DEFINITIONS AND COMPUTATION AIDS FOR THE CALCULUS OF FINITE DIFFERENCES

TABLE A–1

FINITE DIFFERENCE OPERATORS

Boolean shift (displacement) operator,	$E^a f(x) = f(x+a)$
First forward difference, delta,	$\Delta f(x) = f(x+1) - f(x)$
First backward difference, nabla,	$\nabla f(x) = f(x) - f(x-1)$
First forward mean, nu,	$\mathrm{N} f(x) = 1/2[f(x) + f(x+1)]$
First backward mean, un,	$\text{И} f(x) = 1/2[f(x) + f(x-1)]$
First central difference, l.c. delta,	$\delta f(x) = f(x+1/2) - f(x-1/2)$
First central mean, l.c. mu,	$\mu f(x) = 1/2[f(x+1/2) + f(x-1/2)]$
Second central difference, debla,	$\Delta\!\!\!\!\nabla f(x) = f(x+1) - 2f(x) + f(x-1)$
Second mean difference, multa,	$\boxminus f(x) = 1/2[f(x+1) - f(x-1)]$

RELATIONS BETWEEN OPERATORS

$$\Delta = E - 1 = E\nabla = 2(\mathrm{N} - 1) = E^{\frac{1}{2}}\delta$$

$$\nabla = 1 - E^{-1} = E^{-1}\Delta = 2(1 - \text{И}) = E^{-\frac{1}{2}}\delta$$

$$\mathrm{N} = \tfrac{1}{2}(E + 1) = E\text{И} = \tfrac{1}{2}(\Delta + 2) = E^{\frac{1}{2}}\mu$$

$$\text{И} = \tfrac{1}{2}(1 + E^{-1}) = E^{-1}\mathrm{N} = \tfrac{1}{2}(2 - \nabla) = E^{-\frac{1}{2}}\mu$$

$$\Delta\!\!\!\!\nabla = E - 2 + E^{-1} = \Delta - \nabla = \Delta\nabla = \delta^2$$

$$\boxminus = \tfrac{1}{2}(E - E^{-1}) = \tfrac{1}{2}(\Delta + \nabla) = \mathrm{N}\nabla = \text{И}\Delta = \mu\delta = \mathrm{N} - \text{И}$$

$$4\boxminus^2 = \Delta\!\!\!\!\nabla\,[\Delta\!\!\!\!\nabla + 4]$$

$$\Delta\!\!\!\!\nabla + 4 = (2\mathrm{N})(2\text{И})$$

OPERATIONS ON SOLUTIONS TO $[\nabla\!\!\!/ + 2\gamma]\, W(r) = 0$

	$\gamma = 0$		$\gamma = 2$		$\gamma < 0$	$\cosh\lambda = 1-\gamma$	$0 < \gamma < 2$	$\cos\lambda = 1-\gamma$	$\gamma > 2$	$\cosh\lambda = \gamma - 1$
$W(r)$	1	r	$(-1)^r$	$r(-1)^r$	$\cosh\lambda r$	$\sinh\lambda r$	$\cos\lambda r$	$\sin\lambda r$	$(-1)^r\cosh\lambda r$	$(-1)^r\sinh\lambda r$
$\Delta W(r)$	0	1	$[-2]\cdot(-1)^r$	$-[2r+1]\cdot(-1)^r$	$[2\sinh\frac{1}{2}\lambda]\cdot\sinh\lambda(r+\frac{1}{2})$	$[2\sinh\frac{1}{2}\lambda]\cdot\cosh\lambda(r+\frac{1}{2})$	$[-2\sin\frac{1}{2}\lambda]\cdot\sin\lambda(r+\frac{1}{2})$	$[2\sin\frac{1}{2}\lambda]\cdot\cos\lambda(r+\frac{1}{2})$	$[-2\cosh\frac{1}{2}\lambda]\cdot(-1)^r\cosh\lambda(r+\frac{1}{2})$	$[-2\cosh\frac{1}{2}\lambda]\cdot(-1)^r\sinh\lambda(r+\frac{1}{2})$
$\delta W(r)$	0	1	$[2i]\cdot(-1)^r$	$[2i]\cdot r(-1)^r$	$[2\sinh\frac{1}{2}\lambda]\cdot\sinh\lambda r$	$[2\sinh\frac{1}{2}\lambda]\cdot\cosh\lambda r$	$[-2\sin\frac{1}{2}\lambda]\cdot\sin\lambda r$	$[2\sin\frac{1}{2}\lambda]\cdot\cos\lambda r$	$[2i\cosh\frac{1}{2}\lambda]\cdot(-1)^r\cosh\lambda r$	$[2i\cosh\frac{1}{2}\lambda]\cdot(-1)^r\sinh\lambda r$
$\nabla W(r)$	0	1	$[2]\cdot(-1)^r$	$[2r-1]\cdot(-1)^r$	$[2\sinh\frac{1}{2}\lambda]\cdot\sinh\lambda(r-\frac{1}{2})$	$[2\sinh\frac{1}{2}\lambda]\cdot\cosh\lambda(r-\frac{1}{2})$	$[-2\sin\frac{1}{2}\lambda]\cdot\sin\lambda(r-\frac{1}{2})$	$[2\sin\frac{1}{2}\lambda]\cdot\cos\lambda(r-\frac{1}{2})$	$[2\cosh\frac{1}{2}\lambda]\cdot(-1)^r\cosh\lambda(r-\frac{1}{2})$	$[2\cosh\frac{1}{2}\lambda]\cdot(-1)^r\sinh\lambda(r-\frac{1}{2})$
$NW(r)$	1	$r+\frac{1}{2}$	0	$[-\frac{1}{2}]\cdot(-1)^r$	$[\cosh\frac{1}{2}\lambda]\cdot\cosh\lambda(r+\frac{1}{2})$	$[\cosh\frac{1}{2}\lambda]\cdot\sinh\lambda(r+\frac{1}{2})$	$[\cos\frac{1}{2}\lambda]\cdot\cos\lambda(r+\frac{1}{2})$	$[\cos\frac{1}{2}\lambda]\cdot\sin\lambda(r+\frac{1}{2})$	$[-\sinh\frac{1}{2}\lambda]\cdot(-1)^r\sinh\lambda(r+\frac{1}{2})$	$[-\sinh\frac{1}{2}\lambda]\cdot(-1)^r\cosh\lambda(r+\frac{1}{2})$
$\mu W(r)$	1	r	0	$[\frac{1}{2}i]\cdot(-1)^r$	$[\cosh\frac{1}{2}\lambda]\cdot\cosh\lambda r$	$[\cosh\frac{1}{2}\lambda]\cdot\sinh\lambda r$	$[\cos\frac{1}{2}\lambda]\cdot\cos\lambda r$	$[\cos\frac{1}{2}\lambda]\cdot\sin\lambda r$	$[i\sinh\frac{1}{2}\lambda]\cdot(-1)^r\sinh\lambda r$	$[i\sinh\frac{1}{2}\lambda]\cdot(-1)^r\cosh\lambda r$
ИW(r)	1	$r-\frac{1}{2}$	0	$[\frac{1}{2}]\cdot(-1)^r$	$[\cosh\frac{1}{2}\lambda]\cdot\cosh\lambda(r-\frac{1}{2})$	$[\cosh\frac{1}{2}\lambda]\cdot\sinh\lambda(r-\frac{1}{2})$	$[\cos\frac{1}{2}\lambda]\cdot\cos\lambda(r-\frac{1}{2})$	$[\cos\frac{1}{2}\lambda]\cdot\sin\lambda(r-\frac{1}{2})$	$[\sinh\frac{1}{2}\lambda]\cdot(-1)^r\sinh\lambda(r-\frac{1}{2})$	$[\sinh\frac{1}{2}\lambda]\cdot(-1)^r\cosh\lambda(r-\frac{1}{2})$
$[\Delta+\gamma]\,W(r)$	0	1	0	$[-1]\cdot(-1)^r$	$\sinh\lambda$ $\sinh\lambda r$	$\sinh\lambda$ $\cosh\lambda r$	$[-\sin\lambda]\cdot\sin\lambda r$	$[\sin\lambda]\cdot\cos\lambda r$	$[-\sinh\lambda]\cdot(-1)^r\sinh\lambda r$	$[-\sinh\lambda]\cdot(-1)^r\cosh\lambda r$
$[\gamma-\nabla]\,W(r)$	0	-1	0	$(-1)^r$	$[-\sinh\lambda]\cdot\sinh\lambda r$	$[-\sinh\lambda]\cdot\cosh\lambda r$	$[\sin\lambda]\cdot\sin\lambda r$	$[-\sin\lambda]\cdot\cos\lambda r$	$[\sinh\lambda]\cdot(-1)^r\sinh\lambda r$	$[\sinh\lambda]\cdot(-1)^r\cosh\lambda r$
$\nabla\!\!\!/\, W(r)$	0	0	$[-4]\cdot(-1)^r$	$[-4]\cdot r(-1)^r$	$[4\sinh^2\frac{1}{2}\lambda]\cdot\cosh\lambda r$	$[4\sinh^2\frac{1}{2}\lambda]\cdot\sinh\lambda r$	$[-4\sin^2\frac{1}{2}\lambda]\cdot\cos\lambda r$	$[-4\sin^2\frac{1}{2}\lambda]\cdot\sin\lambda r$	$[-4\cosh^2\frac{1}{2}\lambda]\cdot(-1)^r\cosh\lambda r$	$[-4\cosh^2\frac{1}{2}\lambda]\cdot(-1)^r\sinh\lambda r$
$\square\!\!\!/\, W(r)$	0	1	0	$[-1]\cdot(-1)^r$	$[\sinh\lambda]\cdot\sinh\lambda r$	$[\sinh\lambda]\cdot\cosh\lambda r$	$[-\sin\lambda]\cdot\sin\lambda r$	$[\sin\lambda]\cdot\cos\lambda r$	$-[\sinh\lambda]\cdot(-1)^r\sinh\lambda r$	$-[\sinh\lambda]\cdot(-1)^r\cosh\lambda r$

TABLE A-2

HOMOGENEOUS SOLUTIONS TO $[\nabla + 2\gamma] W(r) = 0$ FOR VARIOUS BOUNDARY CONDITIONS

BOUNDARY CONDITIONS	$\gamma < 0$ $\cosh\lambda - 1 - \gamma$	$\gamma = 0$	$0 < \gamma < 2$ $\cos\lambda = 1 - \gamma$	$\gamma = 2$		$\gamma > 2$	$\cosh\lambda = \gamma - 1$
				n - even	n - odd	n - even	n - odd
$W(0) = 1$ $W(n) = 1$	$\frac{\cosh\lambda\left(\frac{n}{2} - r\right)}{\cosh\frac{\lambda n}{2}}$	1	$\frac{\cos\lambda\left(\frac{n}{2} - r\right)}{\cos\frac{\lambda n}{2}}$	$(-1)^r$	$(-1)^r\left(1 - \frac{2r}{n}\right)$	$\frac{(-1)^r\cosh\lambda\left(\frac{n}{2} - r\right)}{\cosh\frac{\lambda n}{2}}$	$\frac{(-1)^r\sinh\lambda\left(\frac{n}{2} - r\right)}{\sinh\frac{\lambda n}{2}}$
$W(0) = 1$ $W(n) = -1$	$\frac{\sinh\lambda\left(\frac{n}{2} - r\right)}{\sinh\frac{\lambda n}{2}}$	$1 - \frac{2r}{n}$	$\frac{\sin\lambda\left(\frac{n}{2} - r\right)}{\sin\frac{\lambda n}{2}}$	$(-1)^r\left(1 - \frac{2r}{n}\right)$	$(-1)^r$	$\frac{(-1)^r\sinh\lambda\left(\frac{n}{2} - r\right)}{\sinh\frac{\lambda n}{2}}$	$\frac{(-1)^r\cosh\lambda\left(\frac{n}{2} - r\right)}{\cosh\frac{\lambda n}{2}}$
$[\Delta + \gamma] W(0) = 1$ $[\gamma - \nabla] W(n) = 1$	$-\frac{\cosh\lambda\left(\frac{n}{2} - r\right)}{\sinh\lambda\sinh\frac{\lambda n}{2}}$		$\frac{\cos\lambda\left(\frac{n}{2} - r\right)}{\sin\lambda\sin\frac{\lambda n}{2}}$			$\frac{(-1)^r\cosh\lambda\left(\frac{n}{2} - r\right)}{\sinh\lambda\sinh\frac{\lambda n}{2}}$	$\frac{(-1)^r\sinh\lambda\left(\frac{n}{2} - r\right)}{\sinh\lambda\cosh\frac{\lambda n}{2}}$
$[\Delta + \gamma] W(0) = 1$ $[\gamma - \nabla] W(n) = -1$	$-\frac{\sinh\lambda\left(\frac{n}{2} - r\right)}{\sinh\lambda\cosh\frac{\lambda n}{2}}$		$-\frac{\sin\lambda\left(\frac{n}{2} - r\right)}{\sin\lambda\cos\frac{\lambda n}{2}}$			$\frac{(-1)^r\sinh\lambda\left(\frac{n}{2} - r\right)}{\sinh\lambda\cosh\frac{\lambda n}{2}}$	$\frac{(-1)^r\cosh\lambda\left(\frac{n}{2} - r\right)}{\sinh\lambda\sinh\frac{\lambda n}{2}}$
$W(0) = 0$ $[\gamma - \nabla] W(n) = 1$	$-\frac{\sinh\lambda r}{\sinh\lambda\cosh\lambda n}$	$-r$	$\frac{\sin\lambda r}{\sin\lambda\cos\lambda n}$	$(-1)^r\, r$	$-(-1)^r\, r$	$\frac{(-1)^r\sinh\lambda r}{\sinh\lambda\cosh\lambda n}$	$-\frac{(-1)^r\sinh\lambda r}{\sinh\lambda\cosh\lambda n}$
$[\Delta + \gamma] W(0) = 0$ $W(n) = 1$	$\frac{\cosh\lambda r}{\cosh\lambda n}$	1	$\frac{\cos\lambda r}{\cos\lambda n}$	$(-1)^r$	$-(-1)^r$	$\frac{(-1)^r\cosh\lambda r}{\cosh\lambda n}$	$-\frac{(-1)^r\cosh\lambda r}{\cosh\lambda n}$
$(\Delta + 2\gamma) W(0) = 1$ $(2\gamma - \nabla) W(n) = 1$	$\frac{-\cosh\lambda\left(\frac{n}{2} - r\right)}{2\sinh\frac{\lambda}{2}\sinh\frac{\lambda}{2}(n+1)}$						
$(\Delta + 2\gamma) W(0) = 1$ $(2\gamma - \nabla) W(n) = -1$	$\frac{-\sinh\lambda\left(\frac{n}{2} - r\right)}{2\sinh\frac{\lambda}{2}\cosh\frac{\lambda}{2}(n+1)}$						

TABLE A - 3

A-2.5

STIRLING NUMBERS OF FIRST KIND[1]
CONVERSION OF FACTORIAL* SERIES TO POWER SERIES

$$\begin{Bmatrix} (r)_1 \\ (r)_2 \\ (r)_3 \\ (r)_4 \\ (r)_5 \\ (r)_6 \end{Bmatrix} = \begin{bmatrix} 1 & 0 & 0 & 0 & 0 & 0 \\ -1 & 1 & 0 & 0 & 0 & 0 \\ 2 & -3 & 1 & 0 & 0 & 0 \\ -6 & 11 & -6 & 1 & 0 & 0 \\ 24 & -50 & 35 & -10 & 1 & 0 \\ -120 & 274 & -225 & 85 & -15 & 1 \end{bmatrix} \begin{Bmatrix} r \\ r^2 \\ r^3 \\ r^4 \\ r^5 \\ r^6 \end{Bmatrix}$$

STIRLING NUMBERS OF SECOND KIND[2]
CONVERSION OF POWER SERIES TO FACTORIAL SERIES

$$\begin{Bmatrix} r \\ r^2 \\ r^3 \\ r^4 \\ r^5 \\ r^6 \end{Bmatrix} = \begin{bmatrix} 1 & 0 & 0 & 0 & 0 & 0 \\ 1 & 1 & 0 & 0 & 0 & 0 \\ 1 & 3 & 1 & 0 & 0 & 0 \\ 1 & 7 & 6 & 1 & 0 & 0 \\ 1 & 15 & 25 & 10 & 1 & 0 \\ 1 & 31 & 90 & 65 & 15 & 1 \end{bmatrix} \begin{Bmatrix} (r)_1 \\ (r)_2 \\ (r)_3 \\ (r)_4 \\ (r)_5 \\ (r)_6 \end{Bmatrix}$$

TABLE A-4

*Factorial $(r)_n = r(r-1)\ (r-2)\ \ldots\ (r-n+2)$ n factors
e.g. $(r)_2 = r(r-1) \equiv r^2 - r$

1. Jordan, "Calculus of Finite Differences", Chelsea, N.Y.,1950, p. 144.

2. ibid., p. 170.

TABLE OF DIFFERENCE OPERATORS ACTING ON $\left(\frac{n}{2} - r\right)^x$

First forward difference Δ

$$\Delta\left(\frac{n}{2} - r\right) = -1$$

$$\Delta\left(\frac{n}{2} - r\right)^2 = -2\left(\frac{n}{2} - r\right) + 1$$

$$\Delta\left(\frac{n}{2} - r\right)^3 = -3\left(\frac{n}{2} - r\right)^2 + 3\left(\frac{n}{2} - r\right) - 1$$

$$\Delta\left(\frac{n}{2} - r\right)^4 = -4\left(\frac{n}{2} - r\right)^3 + 6\left(\frac{n}{2} - r\right)^2 - 4\left(\frac{n}{2} - r\right) + 1$$

$$\Delta\left(\frac{n}{2} - r\right)^5 = -5\left(\frac{n}{2} - r\right)^4 + 10\left(\frac{n}{2} - r\right)^3 - 10\left(\frac{n}{2} - r\right)^2 + 5\left(\frac{n}{2} - r\right) - 1$$

$$\Delta\left(\frac{n}{2} - r\right)^6 = -6\left(\frac{n}{2} - r\right)^5 + 15\left(\frac{n}{2} - r\right)^4 - 20\left(\frac{n}{2} - r\right)^3 + 15\left(\frac{n}{2} - r\right)^2 - 6\left(\frac{n}{2} - r\right) + 1$$

Second central difference $\nabla\!\!\!/$

$$\nabla\!\!\!/\left(\frac{n}{2} - r\right) = 0$$

$$\nabla\!\!\!/\left(\frac{n}{2} - r\right)^2 = 2$$

$$\nabla\!\!\!/\left(\frac{n}{2} - r\right)^3 = 6\left(\frac{n}{2} - r\right)$$

$$\nabla\!\!\!/\left(\frac{n}{2} - r\right)^4 = 12\left(\frac{n}{2} - r\right)^2 + 2$$

$$\nabla\!\!\!/\left(\frac{n}{2} - r\right)^5 = 20\left(\frac{n}{2} - r\right)^3 + 10\left(\frac{n}{2} - r\right)$$

$$\nabla\!\!\!/\left(\frac{n}{2} - r\right)^6 = 30\left(\frac{n}{2} - r\right)^4 + 30\left(\frac{n}{2} - r\right)^2 + 2$$

Mean difference $\boxminus$

$$\boxminus\left(\frac{n}{2} - r\right) = -1$$

$$\boxminus\left(\frac{n}{2} - r\right)^2 = -2\left(\frac{n}{2} - r\right)$$

$$\boxminus\left(\frac{n}{2} - r\right)^3 = -3\left(\frac{n}{2} - r\right)^2 - 1$$

$$\boxminus\left(\frac{n}{2} - r\right)^4 = -4\left(\frac{n}{2} - r\right)^3 - 4\left(\frac{n}{2} - r\right)$$

$$\boxminus\left(\frac{n}{2} - r\right)^5 = -5\left(\frac{n}{2} - r\right)^4 - 10\left(\frac{n}{2} - r\right)^2 - 1$$

$$\boxminus\left(\frac{n}{2} - r\right)^6 = -6\left(\frac{n}{2} - r\right)^5 - 20\left(\frac{n}{2} - r\right)^3 - 6\left(\frac{n}{2} - r\right)$$

Inverse debla

$$\nabla\!\!\!/^{-1} A = \frac{A}{2}\left(\frac{n}{2} - r\right)^2$$

$$\nabla\!\!\!/^{-1} A\left(\frac{n}{2} - r\right) = \frac{A}{6}\left(\frac{n}{2} - r\right)^3$$

$$\nabla\!\!\!/^{-1} A\left(\frac{n}{2} - r\right)^2 = \frac{A}{12}\left[\left(\frac{n}{2} - r\right)^4 - \left(\frac{n}{2} - r\right)^2\right]$$

$$\nabla\!\!\!/^{-1} A\left(\frac{n}{2} - r\right)^3 = A\left[\frac{1}{20}\left(\frac{n}{2} - r\right)^5 - \frac{1}{12}\left(\frac{n}{2} - r\right)^3\right]$$

$$\nabla\!\!\!/^{-1} A\left(\frac{n}{2} - r\right)^4 = A\left[\frac{1}{30}\left(\frac{n}{2} - r\right)^6 - \frac{1}{12}\left(\frac{n}{2} - r\right)^4 + \frac{1}{20}\left(\frac{n}{2} - r\right)^2\right]$$

TABLE A-5

APPENDIX III

FOURIER SERIES

Solutions to the continuous and discrete second order models derived in Lecture II for the fill load condition – Eqs. A-3.1 and A-3.2 below – play a major role in the field analysis of structural units and assemblages.

(A-3.1) $$(D_x^2 + \alpha_i^2)F_i(x) = 0 \qquad \alpha_i^2 > 0$$

(A-3.2) $$(\Delta_r + 2\gamma_k)F_k(r) = 0 \qquad 0 < \gamma_k < 2$$

(A-3.3) $$F_i(x) = \overset{*}{A}_i \sin \alpha_i x + \overset{*}{B}_i \cos \alpha_i x$$

(A-3.4a) $$F_k(r) = A_k \sin \lambda_k r + B_k \cos \lambda_k r$$

(A-3.4b) $$\cos \lambda_k = 1 - \gamma_k$$

The utility of such solutions is based on the fundamental principle that the solution to a linear mathematical model can be written as a linear combination of the solutions to the related Eigen value model. As a result of the uniquely simple recurrence relations that characterize sinuoidal functions,

$$D_x^2 F_i(x) = -\alpha_i^2 F_i(x) \quad \text{and} \quad \Delta_r F_k(r) = -2\gamma_k F_k(r),$$

the class of "related models" amenable to solution in the form of sinusoidal series is quite broad. Specifically, if a governing equation contains only even operators – derivatives or differences – with respect to one of the independent variables, a solution can be written in the form of a sinusoidal series on that variable whose coefficients are functions of the other independent variables (or constants in the case of a single independent variable). Such equations arise frequently in all branches of mathematical physics, especially in structural mechanics. An extensive use is made of sinusoidal series throughout this book, a collection of properties and formulas for their use in structural analysis is summarized in this appendix.

Continuous Variable Series. – To uniquely specify the infinite set of sinusoidal Eigen functions shown in Eq. A-3.3, they must be constrained to satisfy two homogeneous boundary conditions. For example, consider the following arbitrary boundary conditions:

$$(cD_x + d)F_i\,(a) = 0\ ; \quad (c'D_x + d')F_i\,(b) = 0 \qquad \text{(A-3.5)}$$

It can shown that the set of functions, $F_i(x)$, satisfying Eqs. A-3.1 and A-3.5 are orthogonal and can be used to express any continuous function in the range $a < x < b$; that is,

$$\int_a^b F_i\,(x)F_{i'}\,(x)\,dx = \Gamma_i\,\delta_i^{i'} \qquad \text{(A-3.6)}$$

$$f(x) = \sum_{i=0}^{\infty} \overset{*}{C}_i\,F_i\,(x) \qquad \text{(A-3.7)}$$

in which $\delta_i^{i'}$ denotes the Kronecker delta and Γ_i is the normalization factor. These quantities are defined as follows:

$$\delta_i^{i'} = \begin{bmatrix} 0 & i \neq i' \\ & \\ 1 & i = i' \end{bmatrix} ; \quad \Gamma_i = \int_a^b [F_i\,(x)]^2\,dx \qquad \text{(A-3.8a,b)}$$

The orthogonality property, Eq. A-3.6, can be used to derive an explicit formula for the coefficients in the expansion of f(x), Eq. A-3.7, that is,

$$\int_a^b F_{i'}\,(x)f(x)dx = \sum_{i=0}^{\infty} \overset{*}{C}_i\,\Gamma_i\,\delta_i^{i'}$$

$$\overset{*}{C}_i = \frac{1}{\Gamma_i}\int_a^b f(x)\;F_i\,(x)\;dx \qquad \text{(A-3.9)}$$

The general approach outlined in Eqs. A-3.5 – A-3.9 is of theoretical interest but of limited practical computational utility as for arbitrary boundary conditions the Eigen values α_i appear as the roots of a transcendental equation that cannot be solved in closed form. For example, the boundary conditions $F_i\,(0) = 0$ and $(D_x - \sigma)\;F_i\,(a) = 0$ yield the characteristic values as the roots of , $\tan a\alpha_i - \alpha_i = 0$. However, there are two boundary conditions that yield α_i explicitly – symmetry or antisymmetry with respect to the boundary point – resulting in a closed form series. These special cases are directly applicable to a variety of realistic structural boundary conditions and special techniques permit their extension to cover other situations; thus, attention will be restricted to those sets of functions from Eq. A-3.3, for which α_i can be found explicitly.

Infinite Sine Series. – The most commonly used Fourier series is the infinite sine series to represent a continuous function over a finite range. The model for the Eigen values and functions is Eq. A-3.1 plus the boundary conditions shown in Eq. A-3.10 below (resulting in a condition of antisymmetry about $x = 0$ and $x = a$). The orthogonality expression and expansion formula, often termed a half-range expansion, are given by Eqs. A-3.12.

(A-3.10a,b,c) $$F_i(0) = F_i(a) = 0 \qquad \alpha_i = \frac{i\pi}{a}$$

(A-3.11) $$\int_0^a \sin\alpha_i x \sin\alpha_{i'} x \, dx = \frac{a}{2}\delta_i^{i'}$$

(A-3.12a) $$f(x) = \sum_{i=1}^{\infty} \overset{*}{A}_i \sin\alpha_i x \qquad 0 < x < a$$

(A-3.12b) $$\overset{*}{A}_i = \frac{2}{a}\int_0^a f(x) \sin\alpha_i x \, dx$$

A fundamental problem in the study of continuous field models is the determination of a solution for a unit impulse load; that is, when the loading or inhomogeneous term is a Dirac delta distribution, $\delta(x - \xi)$, which has the following properties

$$\delta(x-\xi) = \begin{bmatrix} 0 & x \neq \xi \\ \infty & x = \xi \end{bmatrix} ; \qquad \int_{-\infty}^{+\infty} f(x)\,\delta(x-\xi)\,dx = f(\xi)$$

(A-3.13a)

Thus, from Eq. A-3.12b:

$$\delta(x-\xi) = \frac{2}{a}\sum_{i=1}^{\infty} \sin\alpha_i\xi \sin\alpha_i x \qquad 0 < x,\xi < a$$

(A-3.14)

A second expansion problem frequently encountered in practice is that of expressing an algebraic polynomial in series form. Through judicious use of symmetry and antisymmetry about $x = a/2$ and the characteristic sinusoidal recurrence relation $D_x^2 F_i(x) = -\alpha_i^2 F_i(x)$, this potentially cumbersome problem can be solved almost by inspection. For example, consider the case of a constant,

$f(x) = 1$. Use of Eq. A-3.12b yields

$$1 = \frac{2}{a}\sum_{i=1}^{\infty}\frac{1-\cos i\pi}{\alpha_i}\sin\alpha_i x = \frac{4}{a}\sum_{i=1,3,\ldots}^{\infty}\frac{1}{\alpha_i}\sin\alpha_i x \qquad 0 < x < a \tag{A-3.15}$$

Through consideration of the fact that $f^{\$}(x) = 1/2$ and $f^{a/s}(x) = 1/2(1-2x/a)$ are the symmetric and antisymmetric components of $f(x) = 1 - x/a$, one can, without further computation, write the following expansion for the other linear component.

$$1-2\frac{x}{a} = \frac{4}{a}\sum_{i=2,4,\ldots}^{\infty}\frac{1}{\alpha_i}\sin\alpha_i x \qquad 0 < x < a \tag{A-3.16}$$

(Note that in Eqs. A-3.15 and A-3.16 the series are zero at $x = 0,a$. As a consequence, truncation of the series after a few terms will represent the function poorly at $x = \epsilon$, $a - \epsilon$ as $\epsilon \to 0$). By eliminating the arbitrary rigid body term that results from a double intergration – a term that can be added by use of Eqs. A-3.15 and A-3.16 when required – the sine series representation of higher order algebraic terms can now be written by induction; i.e.,

$$\frac{1}{2}x(a-x) = \frac{4}{a}\sum_{i=1,3,\ldots}^{\infty}\frac{1}{\alpha_i^{\,3}}\sin\alpha_i x \qquad 0 \leqslant x \leqslant a \tag{A-3.17a}$$

$$\frac{1}{2}x(a-x) \equiv \frac{a^2}{8}\left[1-\left(1-2\frac{x}{a}\right)^2\right] \tag{A-3.17b}$$

$$\frac{1}{6}x(a-x)\left(1-2\frac{x}{a}\right) = \frac{4}{a}\sum_{i=2,4,\ldots}^{\infty}\frac{1}{\alpha_i^{\,3}}\sin\alpha_i x \qquad 0 \leqslant x \leqslant a \tag{A-3.18a}$$

$$\frac{1}{6}x(a-x)\left(1-2\frac{x}{a}\right) \equiv \frac{a^2}{24}\left[\left(1-2\frac{x}{a}\right)-\left(1-2\frac{x}{a}\right)^3\right] \tag{A-3.18b}$$

etc. for terms of fourth degree and higher.

In order to illustrate use of these inductively derived formulae for the expansion of algebraic functions, consider the following two examples:

$$f(x) = A + Cx^2 = f^{\$}(x) + f^{a/s}(x) \tag{A-3.19a}$$

(A-3.19b) $g(x) \;=\; Bx + Dx^3 = g^{\$}(x) + g^{a/s}(x)$

(A.3.19c) $f^{\$}(x) = \left(A + C\frac{a^2}{4}\right) + C\left(\frac{a}{2} - x\right)^2 = \left(A + C\frac{a^2}{2}\right) - Cx(a-x)$

(A-3.19d) $f^{\$}(x) = \frac{4}{a}\sum_{i=1,3,\dots}^{\infty}\left[\frac{A + C\frac{a^2}{2}}{\alpha_i} - \frac{2C}{\alpha_i^3}\right]\sin\alpha_i x$

(A-3.19e) $f^{a/s}(x) = -Ca\left(\frac{a}{2} - x\right) = -\frac{4}{a}\sum_{i=2,4,\dots}^{\infty}\frac{C\frac{a^2}{2}}{\alpha_i}\sin\alpha_i x$

(A-3.19f) $g^{\$}(x) = \frac{a}{2}\left(B + D\frac{a^2}{4}\right) + 3D\frac{a}{2}\left(\frac{a}{2} - x\right)^2$

(A-3.19g) $g^{\$}(x) = 2\sum_{i=1,3,\dots}^{\infty}\left[\frac{B + Da^2}{\alpha_i} - \frac{6D}{\alpha_i^3}\right]\sin\alpha_i x$

(A-3.19h) $g^{a/s}(x) = -\left(\frac{a}{2} - x\right)\left[B + 3\frac{a^2}{4}D + D\left(\frac{a}{2} - x\right)^2\right]$

(A-3.19i) $g^{a/s}(x) = -2\sum_{i=2,4,\dots}^{\infty}\left[\frac{B + Da^2}{\alpha_i} - \frac{6D}{\alpha_i^3}\right]\sin\alpha_i x$

Note that series for higher degree algebraic terms, which have higher powers of i in the denominator as a result of integration of the Fourier series, converges more rapidly. In using such series to solve a differential equation, the solution series usually converges more rapidly than does the loading series, which may even be divergent as in the case of the impulse loading, Eq. A-3.14.

Infinite Cosine Series – The second major Fourier series required for the representation of continuous functions is the set of Eigen functions satisfying a condition of symmetry about the boundary points; that is,

(A-3.20a) $F_i(x) = B_i^* \cos\alpha_i x$

(A-3.20b,c,d) $D_x F_i(0) = D_x F_i(a) = 0 \qquad \alpha_i = \frac{i\pi}{a}$

A distinct, and ocassionally troublesome, feature of the cosine series is its representation of the constant term in the rigid body solution to Eq. 3.1 through inclusion of the term for $i = 0$. The orthogonality expression and expansion formula are given by Eqs. A-3.21 and A-3.22 below.

$$\int_0^a \cos \alpha_i x \cos \alpha_{i'} x \, dx = \frac{a}{2\phi_i} \delta_i^{i'} \qquad \text{(A-3.21a)}$$

$$\phi_i = 1 - \frac{1}{2} \delta_i^0 \qquad \text{(A-3.21b)}$$

$$f(x) = \sum_{i=0}^{\infty} \overset{*}{B}_i \cos \alpha_i x \qquad 0 \leqslant x \leqslant a \qquad \text{(A-3.22a)}$$

$$\overset{*}{B}_i = \frac{2\phi_i}{a} \int_0^a f(x) \cos \alpha_i x \, dx \qquad \text{(A-3.22b)}$$

The cosine expansion formula may be written in an alternate from by including the normalization weighting function, ϕ_i , as a factor in the series coefficients; i.e.,

$$f(x) = \sum_{i=0}^{\infty} \phi_i \overset{*}{B}{}'_i \cos \alpha_i x \qquad 0 \leqslant x \leqslant a \qquad \text{(A-3.23a)}$$

$$\overset{*}{B}{}'_i = \frac{2}{a} \int_0^a f(x) \cos \alpha_i x \, dx \qquad \text{(A-3.23b)}$$

The two forms, Eqs. 3.22 and 3.23 are used interchangeably here. Expansion of the continuous impulse function is:

$$\delta(x - \xi) = \frac{2}{a} \sum_{i=0}^{\infty} \phi_i \cos \alpha_i \xi \cos \alpha_i x \qquad 0 \leqslant x, \xi \leqslant 0 \qquad \text{(A-3.24)}$$

The inductive method of expanding algebraic polynomials (see previous section) is also applicable here, with the difference that for i odd the cosine terms are antisymmetric about $x = a/2$ whereas the corresponding sine terms were symmetric. Use of A-3.12b (or A-3.23b) to find the cosine expansion of a constant yields the single terms series, actually an identity, given by Eq. A-3.25a below; however, it is useful to have available an alternate divergent infinite series (Eq. A-3.25b) – obtained by differentiating Eq. A-3.16. A cosine expansion for the first degree antisymmetric algebraic term is found by use of the standard expansion formula, Eq. A-3.22b or by differentiating the sine series for the second degree symmetric algebraic term, Eq. A-3.17a. The result, by either method, is given by Eq. A-3.26 below.

(A-3.25a) $$1 = \sum_{i=0}^{\infty} \delta_i^0 \cos \alpha_i x$$

(A-3.25b) $$1 = -2 \sum_{i=2,4,\ldots}^{\infty} \cos \alpha_i x$$

(A-3.26) $$1 - 2\frac{x}{a} = \frac{8}{a^2} \sum_{i=1,3,\ldots}^{\infty} \frac{1}{\alpha_i^2} \cos \alpha_i x$$

Expansion for the second degree symmetric algebraic term, except for the rigid body coefficient, is obtained by induction, double integration of Eq. A-3.25b. The rigid body term is easily found by use of Eq. A-3.22b for i = 0 and given by Eq. A-3.27 below.
(The result can also be found through differentiation of the sine series for the third degree antisymmetric algebraic term, Eq. A-3.18a.) Similarly, the cosine series for the third degree antisymmetric algebraic term, is written by induction, double integration of Eq. A-3.26, as shown in Eq. A-3.28 below.

(A-3.27) $$\frac{1}{2}\left(1 - 2\frac{x}{a}\right)^2 = \frac{1}{6} + \frac{8}{a^2} \sum_{i=2,4,\ldots}^{\infty} \frac{1}{\alpha_i^2} \cos \alpha_i x$$

(A-3.28) $$3\left(1 - 2\frac{x}{a}\right) - \left(1 - 2\frac{x}{a}\right)^3 = \frac{192}{a^4} \sum_{i=1,3,\ldots}^{\infty} \frac{1}{\alpha_i^4} \cos \alpha_i x$$

Eqs. A-3.25 thru A-3.28 can be used to express the algebraic functions f(x) and g(x) – Eqs. A-3.19 of the previous section – with the following results:

(A-3.29a) $$f^{\$}(x) = A + \frac{1}{3} C a^2 + \sum_{i=2,4,\ldots}^{\infty} \frac{4C}{\alpha_i^2} \cos \alpha_i x$$

(A-3.29b) $$f^{a/s}(x) = -\sum_{i=1,3,\ldots}^{\infty} \frac{4C}{\alpha_i^2} \cos \alpha_i x$$

(A-3.29c) $$g^{\$}(x) = \left(B + \frac{1}{2} D a^2\right)\frac{a}{2} + \sum_{i=2,4,\ldots}^{\infty} \frac{6Da}{\alpha_i^2} \cos \alpha_i x$$

(A-3.29d) $$g^{a/s}(x) = -\frac{4}{a} \sum_{i=1,3,\ldots}^{\infty} \left[\frac{B + \frac{3}{2} D a^2}{\alpha_i^2} - \frac{6D}{\alpha_i^4}\right] \cos \alpha_i x$$

A comparison of the cosine and sine series shows that: for antisymmetric algebraic terms, the cosine series is more rapidly convergent; the cosine series reproduces the expanded function at the points but with zero boundary slopes; and the higher degree symmetric and antisymmetric algebraic terms for which the series were derived inductively have slightly different forms (e.g., compare Eq. A-3.17a with Eq. A-3.27) due to different natural boundary conditions between the terms of the series.

Infinite Mixed Boundary Series. – A full treatment of the infinite Fourier series available for closed form expansions requires consideration of Eigen functions that are antisymmetric with respect to one boundary and symmetric with respect to the other. These two additional series are as follows:

$$F_i(0) = D_x F_i(a) = 0 \qquad F_i(x) = \overset{*}{A_i} \sin \alpha'_i x \qquad \text{(A-3.30a)}$$

$$\alpha'_i = \frac{i\pi}{2a} \qquad i = 1, (2), \infty$$

$$f(x) = \sum_{i=1,3,\ldots}^{\infty} \overset{*}{A_i} \sin \alpha'_i x \;\; ; \;\; \overset{*}{A_i} = \frac{2}{a}\int_0^a f(x) \sin \alpha'_i x \qquad \text{(A-3.30b)}$$

$$D_x F_i(0) = F_i(a) = 0 \qquad F_i(x) = \overset{*}{B_i} \cos \alpha'_i x \qquad \text{(A-3.31a)}$$

$$\alpha'_i = \frac{i\pi}{2a} \qquad i = 1, (2), \infty$$

$$f(x) = \sum_{i=1,3,\ldots}^{\infty} \overset{*}{B_i} \cos \alpha'_i x \qquad \overset{*}{B_i} = \frac{2}{a}\int_0^a f(x) \cos \alpha'_i x \qquad \text{(A-3.31b)}$$

Thus, it may be seen that the two mixed boundary cases can be treated as special cases of the regular sine and cosine series; that is, the case $F_i(0) = D_x F_i(a) = 0$ coincides with the case $F_i(0) = F_i(a) = 0$ (regular sine series) if $a' = 2a$ with symmetry about $x = a'/2$. Similarly, the case $D_x F_i(0) = D_x F_i(a) = 0$ (regular cosine series) if $a' = 2a$ with antisymmetry about $x = a'/2$. These mixed cases are sometimes referred to as quarter range series but the above view (i.e., their interpretation as modified half range series) is sufficient and less susceptible to error.

Discrete Variable Series. – The problem of expanding the function of a discrete variable into a finite sinusoidal series is broader and more fundamental than the analogous continuous variable problem which could have been covered as a limiting case. The arbitrary interior boundary conditions shown in Eq. A-3.32 below and the governing equation, Eq. A-3.2, form the mathematical model to be satisfied by the Eigen functions $F_k(r)$ (or $F(\lambda_k r)$); i.e., the functions in Eq. A-3.4 must be constrained to satisfy two homogeneous boundary conditions in order to uniquely specify the parameter γ_k and the primary range of k. Here, too, one can show that the functions are orthogonal, with respect to a boundary weighting function, and can be used to express any discrete functions in the specified range; e.g., r = 0,(1),m. The general conditions are as follows:

(A-3.32) $$(c\,\Delta_r + d)F_k\,(0) = (d' - c'\,\nabla_r)F_k\,(m) = 0$$

(A-3.33a) $$\sum_{r=0}^{m} \omega_r F_k\,(r) F_{k'}\,(r) = \Gamma_k \delta_k^{k'}$$

(A-3.33b) $$\Gamma_k = \sum_{r=0}^{m} \omega_r [F_k\,(r)]^2$$

(A-3.34a) $$f(r) = \sum_{k=0}^{m} C_k F_k\,(r)$$

(A-3.34b) $$C_k = \frac{1}{\Gamma_k} \sum_{r=0}^{m} \omega_r f(r) F_k\,(r)$$

The inclusion of a weighting function, ω_r, in the orthogonality statement, Eq. A-3.33a, requires an explanation as the governing equation, Eq. A-3.2, has constant coefficients. In conformity with classical analysis, the condition $\omega_r = 1$ is valid for the interior nodes or the range of applicability of the governing equation (i.e., r = 1,(1),m – 1) and the weighting function can be omitted from the orthogonality equations if the range is over the interior nodes only. More generally, any set of solutions whose boundary conditions can be written in a form independent of the equation parameters, γ_k , will satisfy an orthogonality equation with a constant or unit weighting function. On the other hand, some important cases (e.g., the cosine series) require inclusion of the boundary nodes in the orthogonality equation in such a fashion that $\omega_r \neq 1$ on the boundary. Thus, the general conditions must be written with a weighting function to cover such special boundary conditions.

As in the limiting continuous variable case, arbitrary boundary constraints on the discrete Eigen functions, Eq. A-3.4, do not yield the characteristic values, γ_k or λ_k , explicitly but rather as the roots of a transcendental equation. A series of such functions would have limited use; thus, attention will be restricted to those sets of functions for which the characteristic values may be found in closed form. There are four basic boundary conditions (i.e., sixteen possible two boundary combinations) that yield γ_k and λ_k explicitly – antisymmetry or symmetry with respect to the boundary node and antisymmetry or symmetry with respect to a point midway between the boundary and first interior nodes. Finite series of sinusoidal functions that satisfy these special boundary conditions play a primary role in discrete field analysis and a presentation of the formulas required for their use follows. (For convenient reference, each of the four basic boundary conditions will be denoted by a single letter; thus, a discrete Eigen function that is antisymmetric with respect to the initial node and symmetric with respect to the terminal node, is referred to as case a-d.)

Finite Sine Series (Case a-a). – The most frequently used discrete variable Fourier series is the sine series, each term of which is antisymmetric with respect to both the initial and terminal nodes. The boundary conditions, characteristic values, orthogonality equation and expansion formulas are as follows:

$$F_k(0) = F_k(m) = 0 \qquad \lambda_k = \frac{k\pi}{m} \tag{A-3.35a}$$

$$\sum_{r=1}^{m-1} \sin\lambda_k r \,\sin\lambda_{k'} r = \frac{m}{2}\,\delta_k^{k'} \tag{A-3.35b}$$

$$f(r) = \sum_{k=1}^{m-1} A_k \sin\lambda_k r \qquad 0 < r < m \tag{A-3.35c}$$

$$A_k = \frac{2}{m}\sum_{r=1}^{m-1} f(r)\sin\lambda_k r \qquad \text{or} \tag{A-3.35d}$$

$$A_k = \frac{2}{m}\,\Delta_r^{-1}\left[f(r)\sin\lambda_k r\right]\Big|_1^m \tag{A-3.35e}$$

As with continuous fields, a fundamental problem is that of finding a solution for a unit discrete impulse load; that is, for a Kronecker delta load term, defined in Eq. A-3.8a. By use of Eq. A-3.35d the equivalent finite sine series is found to be:

$$\delta_r^{\alpha} = \frac{2}{m}\sum_{k=1}^{m-1} \sin\lambda_k\alpha \,\sin\lambda_k r \qquad 0 < \alpha, r < m \tag{A-3.36}$$

Here too, the problem of expanding an algebraic polynomial is much simplified by use of symmetry and antisymmetry about the range midpoint, $r = m/2$, and the function's recurrance relation, $\nabla_r F_k(r) = -2\gamma_k F_k(r)$. The symmetric component (obtained by use of Eq. A-3.35e) and antisymmetric component (written by induction from the result for the symmetric component) of the first degree polynomial are as follows:

(A-3.37a) $$1 = \frac{2}{m} \sum_{k=1,3,\ldots}^{m-1} \frac{\sin \lambda_k}{\gamma_k} \sin \lambda_k r \equiv \frac{2}{m} \sum_{k=1,3,\ldots}^{m-1} \cot \frac{1}{2} \lambda_k \sin \lambda_k r$$

(A-3.37b) $$1 - 2\frac{r}{m} = \frac{2}{m} \sum_{k=2,4,\ldots}^{m-1} \cot \frac{1}{2} \lambda_k \sin \lambda_k r \qquad 0 < r < m$$

In comparing Eqs. A-3.37 with Eqs. A-3.15 and 3.16, it is noted that the coefficients here, $\cot 1/2\, \lambda_k$, lack the index that causes convergence in the continuous case. Convergence is of little theoretical interest for a finite series, but it may be practical to truncate the finite series if the field is large, say $m > 5$. For this reqson it is helpful to make use of the fact that $\cot 1/2\lambda_k$, resembles a negative power curve in the range $0 < k < m$.

The symmetric second degree polynomial and the antisymmetric third degree polynomial are found by inverse operation by ∇_r on Eqs. A-3.37, as follows:

$$\frac{1}{2} r(m-r) = \frac{1}{m} \sum_{k=1,3,\ldots}^{m-1} \frac{\sin \lambda_k}{\gamma_k^2} \sin \lambda_k r \qquad 0 \leqslant r \leqslant m$$
(A-3.38a)

$$\frac{1}{6} r(m-r)\left(1 - 2\frac{r}{m}\right) = \frac{1}{m} \sum_{k=2,4,\ldots}^{m-1} \frac{\sin \lambda_k}{\gamma_k^2} \sin \lambda_k r \qquad 0 \leqslant r \leqslant m$$
(A-3.38b)

The finite sine series for fourth and higher degree polynomials can be written by repeating the inverse operation. Use of the above series to express arbitrary polynomials – Eqs. A-3.19 with x replaced by r – is illustrated below:

$$f^{\$}(r) + f^{a/s}(r) = A + Cr^2 \; ; \quad g^{\$}(r) + g^{a/s}(r) = Br + Dr^3$$
(A-3.39a,b)

$$f^{\$}(r) = \frac{2}{m}\sum_{k=1,3,\ldots}^{m-1}\left[\frac{A + \frac{1}{2}Cm^2}{\gamma_k} - \frac{C}{\gamma_k^2}\right] \sin\lambda_k \sin\lambda_k r \qquad \text{(A-3.39c)}$$

$$f^{a/s}(r) = \frac{-2}{m}\sum_{k=2,4,\ldots}^{m-1}\left[\frac{\frac{1}{2}Cm^2}{\gamma_k}\right] \sin\lambda_k \sin\lambda_k r \qquad \text{(A-3.39d)}$$

$$g^{\$}(r) = \sum_{k=1,3,\ldots}^{m-1}\left[\frac{B + Dm^2}{\gamma_k} - \frac{3D}{\gamma_k^2}\right] \sin\lambda_k \sin\lambda_k r \qquad \text{(A-3.39e)}$$

$$g^{a/s}(r) = -\sum_{k=2,4,\ldots}^{m-1}\left[\frac{B + Dm^2}{\gamma_k} - \frac{3D}{\gamma_k^2}\right] \sin\lambda_k \sin\lambda_k r \qquad \text{(A-3.39f)}$$

Finite Cosine Series (case d-d). – The cosine series equals the sine series in usefulness as a tool for discrete field analysis. Its application, however, requires more vare due to its inclusion of terms for k = 0,m and the special boundary conditions required for symmetry with espect to the initial and terminal boundary nodes. The boundary conditions, characteristic values, orthogonality equation and expansion formulas are as follows:

$$(\Delta_r + \gamma_k)F_k(0) = (\gamma_k - \nabla_r)F_k(m) = 0 \qquad \text{(A-3.40a)}$$

$$\lambda_k = \frac{k\pi}{m} \quad \omega_r = 1 - \frac{1}{2}(\delta_r^0 + \delta_r^m) \qquad \text{(A-3.40b)}$$

$$\sum_{r=0}^{m} \omega_r \cos\lambda_k r \cos\lambda_{k'} r = \frac{m}{2\omega_k}\delta_k^{k'} \qquad \text{(A-3.40c)}$$

$$f(r) = \sum_{k=0}^{m} B_k \cos\lambda_k r \qquad 0 \leqslant r \leqslant m \qquad \text{(A-3.40d)}$$

$$B_k = \frac{2\omega_k}{m}\sum_{r=0}^{m}\omega_r f(r) \cos\lambda_k r \qquad \text{(A-3.40e)}$$

Note that both boundary statements, Eqs. A-3.40a, as well as the model for the interior nodes, Eq. A-3.2, contain the parameter γ_k . thus there are m + 1 distinct characteristic values for γ_k, and λ_k with as many terms in the series A-3.40d, which permits f(r) to be represented over a larger range; i.e., r = 0,(1),m.

The added values of γ_k are $\gamma_0 = 0$ and $\gamma_m = 2$, which represent degenerate cases in the governing Eq. A-3.2, and the corresponding series terms are B_0 and $B_m(-1)^r$.

Expansion of the discrete impulse function and inductive derivation of representative basic algebraic polynomials are shown below.

(A-3.41a) $$\delta_r^\alpha = \frac{2}{m}\sum_{k=0}^{m} \omega_k \omega_\alpha \cos\lambda_k\alpha \cos\lambda_k r \qquad 0 \leqslant r,\alpha, \leqslant m$$

(A-3.41b) $$\delta_r^0 \pm \delta_r^m = \frac{2}{m}\sum_{\substack{k=0,2,\ldots \\ k=1,3,\ldots}}^{m} \omega_k \cos\lambda_k r$$

(A-3.42a) $$1 = \sum_{k=0}^{m} \delta_k^0 \cos\lambda_k r$$

or

(A-3.42b) $$1 = -2\sum_{k=2,4,\ldots}^{m} \omega_k \cos\lambda_k r \qquad 0 < r < m$$

(A-3.42c) $$1 - 2\frac{r}{m} = \frac{4}{m^2}\sum_{k=1,3,\ldots}^{m} \frac{\omega_k}{\gamma_k} \cos\lambda_k r \qquad 0 \leqslant r \leqslant m$$

(A-3.42d) $$\frac{1}{2}\left(1 - 2\frac{r}{m}\right)^2 = \frac{m^2+2}{6m^2} + \frac{4}{m^2}\sum_{k=2,4,\ldots}^{m} \frac{\omega_k}{\gamma_k} \cos\lambda_k r \quad 0 \leqslant r \leqslant m$$

(A-3.42e) $$\frac{1}{6}\left(1 - 2\frac{r}{m}\right)\left[\frac{m^2}{2} + 1 + r(m-r)\right] = \frac{2}{m^2}\sum_{k=1,3,\ldots}^{m} \frac{\omega_k}{\gamma_k^2} \cos\lambda_k r$$

Equations A-3.41 and A-3.42a,c were obtained by use of the basic coefficient formula, Eq. A-3.40e, while Eqs. A-3.42d,e were found by operating ∇_r^{-1} on Eqs. A-3.42b,c. Equation A-3.42b was obtained by solving for the constant term ($k = 0$) in Eq. A-3.41b; the expression is invalid on the boundary. It should be noted that operation by Δ_r on the cosine series, Eqs. A-3.42, yields the sine series for the algebraic term of the next lower degree but that a similar operation on the sine series yields a cosine series which differes by a boundary function, Eq. A-3.41b, from the cosine series for the algebraic term of the next lower degree. Here, too, the basic algebraic terms were constrained to fit the boundary conditions natural to the corresponding Eigen functions; i.e., for the sine series $F(0) = 0$ whereas for the cosine series the constraint was $\Delta_r F(0) = 1/2\ \nabla_r F(0)$.

Equations A-3.42 can be used to express the arbitrary functions considered in the previous section with the following results:

$$f^{\$}(r) = A + \frac{1}{6}\, C(2m^2 + 1) + \sum_{k=2,4,\ldots}^{m} \frac{2\omega_k C}{\gamma_k} \cos \lambda_k r \qquad \text{(A-3.43a)}$$

$$f^{a/s}(r) = -\sum_{k=1,3,\ldots}^{m} \frac{2\omega_k C}{\gamma_k} \cos \lambda_k r \qquad \text{(A-3.43b)}$$

$$g^{\$}(r) = \frac{m}{2}\left[B + \frac{1}{2}\, D(m^2 + 1)\right] + \frac{m}{2} \sum_{k=2,4,\ldots}^{m} \frac{6D\omega_k}{\gamma_k} \cos \lambda_k r \qquad \text{(A-3.43c)}$$

$$g^{a/s}(r) = -\frac{2}{m} \sum_{k=1,3,\ldots}^{m} \left[\frac{B + D\left(\frac{3}{2}\, m^2 + 1\right)}{\gamma_k} - \frac{3D}{\gamma_k^2}\right] \omega_k \cos \lambda_k r \qquad \text{(A-3.43d)}$$

Displaced Sine Series (Case c-c). – Two of the four basic types of discrete variable Eigen functions do not have continuous variable counterparts. Consider, for example, the set of $F_k(r)$, Eq. A-3.4a, constrained to be antisymmetric about points displaced one half a node space from both the initial and terminal boundary nodes; that is,

$$\not{M}_r F_k(0) = \not{M}_r F_k(m) = 0 \qquad \left(a/s \text{ at } r = -\frac{1}{2},\ m - \frac{1}{2}\right) \qquad \text{(A-3.44a)}$$

$$F_k(r) = \sin \lambda_k \left(r + \frac{1}{2}\right); \quad \lambda_k = \frac{k\pi}{m}; \quad k = 1,(1),m \qquad \text{(A-3.44b)}$$

$$\sum_{r=0}^{m-1} \sin \lambda_k \left(r + \frac{1}{2}\right) \sin \lambda_{k'} \left(r + \frac{1}{2}\right) = \frac{m}{2\omega_k}\, \delta_k^{k'} \qquad \text{(A-3.44c)}$$

$$f(r) = \sum_{k=1}^{m} A_k \sin \lambda_k \left(r + \frac{1}{2}\right) \qquad 0 \leqslant r \leqslant m \qquad \text{(A-3.44d)}$$

$$A_k = \frac{2\omega_k}{m} \sum_{r=0}^{m-1} f(r) \sin \lambda_k \left(r + \frac{1}{2}\right) \qquad \text{(A-3.44e)}$$

in which ω_k is as defined in Eq. A-3.40b but ω_0 is not required here. It should be noted that the form of this set of functions depends upon the location of the points of antisymmetry in relation to the nodes $r = 0,m$. For example, one may require that the second condition in Eq. A-3.44a be replaced by $N_r F_k(m) = 0$ for a/s about $r = -1/2$ and $m + 1/2$. In this case one would replace m by $m + 1$ in Eqs. A-3.44 for $\lambda_k = k\pi/(m+1)$ etc. Adjustments are easily made for such deviations so that it is not necessary to catalogue all the various possibilities. The authors selected the form shown in Eqs. A-3.44 due to its conformability with the series for cases a-a and d-d; i.e., identical λ_k for all series – a necessity in dealing with vector fields for which the components are represented by different Eigen functions.

Displaced Cosine Series (Case b-b). – The fourth basic type of finite series consists of a set of the Eigen functions, $F_k(r)$in Eq. A-3.4a, constrained to conditions of symmetry about half space points adjacent to the boundary nodes; that is,

$$\nabla_r F_k(0) = \nabla_r F_k(m) = 0 \qquad \left(\$ \text{ at } r = -\frac{1}{2},\ m - \frac{1}{2}\right) \tag{A-3.45a}$$

$$F_k(r) = \cos \lambda_k \left(r + \frac{1}{2}\right);\quad \lambda_k = \frac{k\pi}{m},\quad k = 0,(1),m-1 \tag{A-3.45b}$$

$$\sum_{r=0}^{m-1} \cos \lambda_k \left(r + \frac{1}{2}\right) \cos \lambda_{k'} \left(r + \frac{1}{2}\right) = \frac{m}{2\omega_k}\, \delta_k^{k'} \tag{A-3.45c}$$

$$f(r) = \sum_{k=0}^{m-1} B_k \cos \lambda_k \left(r + \frac{1}{2}\right) \qquad 0 \leqslant r \leqslant m \tag{A-3.45d}$$

$$B_k = \frac{2\omega_k}{m} \sum_{r=0}^{m-1} f(r) \cos \lambda_k \left(r + \frac{1}{2}\right) \tag{A-3.45e}$$

in which ω_k is again defined as in Eq. A-3.40b, but here ω_m is not required. As with Case c-c, special conditions may require some adjustments in Eqs. A-3.45. A common variation is the case of a structural latice whose boundaries are free of external forces, which requires symmetry about $r = -1/2$ and $m + 1/2$; thus, one would replace m by $m + 1$ for $\gamma_k = k\pi/(m + 1)$, etc.

Finite Mixed Boundary Series. – In addition to the four basic discrete variable series discussed in the preceding paragraphs there are twelve other combinations of boundary conditions that can be satisfied by sinusoidal Eigen functions whose characteristic values are known in closed form. These additional conditions consist of the "off diagonal" combinations of the types a,b,c and d; e.g., Case a-d (a/s at $r = 0$ and \$ at $r = m$), Case a-b (a/s at $r = 0$ and \$ at $r = m - 1/2$), and Case b-c (\$

at r = − 1/2 and a/s at r = m − 1/2).

All of the finite mixed boundary series can be constructed by doubling the field length of one of the basic four cases and selecting series terms so as to create the appropriate conditions, symmetry or antisymmetry, at the mid field point, which comprises the terminal boundary for the mixed boundary case. For example, such adjustments to handle the above three mixed cases give the following results:

Case a-d

$$F_k(0) = (\gamma_k - \nabla_r)F_k(m) = 0 \qquad \text{(A-3.46a)}$$

$$f(r) = \sum_{k=1,3,\ldots}^{2m-1} A_k \sin\frac{k\pi r}{2m} \qquad 0 < r \leqslant m \qquad \text{(A-3.46b)}$$

$$A_k = \frac{2}{m}\sum_{r=1}^{m} \omega_r\, f(r) \sin\frac{k\pi r}{2m} \qquad \text{(A-3.46c)}$$

Case a-b

$$F_k(0) = \nabla_r F_k(m) = 0 \qquad \text{(A-3.47a)}$$

$$f(r) = \sum_{k=1,3,\ldots}^{2m-2} A_k \sin\frac{k\pi r}{2m-1} \qquad 0 < r < m \qquad \text{(A-3.47b)}$$

$$A_k = \frac{4}{2m-1}\sum_{r=1}^{m-1} A_k \sin\frac{k\pi r}{2m-1} \qquad \text{(A-3.47c)}$$

Case b-c

$$\nabla_r F_k(0) = \mathit{\Lambda}_r F_k(m) = 0 \qquad \text{(A-3.48a)}$$

$$f(r) = \sum_{k=1,3,\ldots}^{2m-2} B_k \cos\frac{k\pi}{2m-1}\left(r+\frac{1}{2}\right) \quad 0 \leqslant r < m \qquad \text{(A-3.48b)}$$

$$B_k = \frac{4}{2m-1}\sum_{r=0}^{m-1} f(r) \cos\frac{k\pi}{2m-1}\left(r+\frac{1}{2}\right) \qquad \text{(A-3.48c)}$$

Expansion formulas for the other nine mixed boundary series, i.e., cases a-c, b-a, b-d, c-a, c-b, c-d, d-a, d-b, and d-c, can be written in a similar fashion.

Modification of Basic Series. – In discrete field analysis, one has more flexibility in selecting a mathematical model than in continuous field analysis. One disadvantage of this flexibility is that the analyst may fail to recognize that his model can be rewritten in a form for which a solution is readily available. On the other hand, one can often utilize the wider choice of model forms to facilitate solution of an otherwise intractable problem. As examples that illustrate possible recognition difficulties, consider the following terminal boundary conditions:

1) $(1 - \nabla_r)F(m') = 0,$

2) $(2\gamma - \nabla_r)F(m') = 0$ and

3) $(2\gamma - 2 - \nabla_r)F(m') = 0.$

Inspection of the boundary operator and consideration of the difference between internal and external boundary statements reveals that these three conditions are identical, respectively, to the following standard conditions:

Case a for $m' = m - 1$,

Case b for $m' = m + 1$, and

Case c for $m' = m + 1$.

For examples that illustrate use of the flexible relation between physical and mathematical boundaries, consider the following cases where pseudo mathematical nodes were added to produce nonstandard conditions at the terminal node of the physical system:

1) Case a at $r = m + 1$,

2) Case d at $r = m + 1$, and

3) Case a at $r = m + n'$.

The resulting conditions at $r = m$, respectively are :

1) $(2\gamma - 1 - \nabla_r)F(m) = 0,$

2) $\left[\frac{\gamma(2\gamma - 3}{\gamma - 1} - \nabla_r\right]F(m) = 0,$

3) $[\gamma + \sinh\lambda \coth\lambda n' - \nabla_r]F(m) = 0,$ in which $\cosh\lambda = \gamma - 1.$

Multiple Series. – Many of the significant field models are multidimensional; that is, they contain more than one independent variable. These independent variables, or coordinates, may be continuous, discrete or mixed. The solutions for such models are frequently written in the form of a multiple sinusoidal series, which usually necessitates the expansion of a multidimensional loading term. This can be done routinely through repeated use of the formulas given here; for example, consider the illustrations below for three classes of two-dimensional functions both variables continuous, both variables discrete, and one variable continuous and the other discrete.

$$f(x,y) = \sum_{j=1}^{\infty} \sum_{i=1}^{\infty} \overset{*}{A}_{ij} \sin \frac{i\pi x}{a} \sin \frac{j\pi y}{b} \qquad \text{(A-3.49a)}$$

$$\overset{*}{A}_{ij} = \frac{4}{ab} \int_0^a \int_0^b f(x,y) \sin \frac{i\pi x}{a} \sin \frac{j\pi y}{b} \, dx \, dy \qquad \text{(A-3.49b)}$$

$$f(r,s) = \sum_{\ell=1}^{n} \sum_{k=0}^{m} B_{k\ell} \cos \frac{k\pi r}{m} \sin \frac{\ell\pi s}{n} \qquad \text{(A-3.50a)}$$

$$B_{k\ell} = \frac{4\omega_k}{mn} \sum_{s=1}^{n-1} \sum_{r=0}^{m} \omega_r \, f(r,s) \cos \frac{k\pi r}{m} \sin \frac{\ell\pi s}{n} \qquad \text{(A-3.50b)}$$

$$f(r,y) = \sum_{j=0}^{\infty} \sum_{k=0}^{m-1} \overset{*}{B}_{kj} \cos \frac{k\pi}{m} \left(r + \frac{1}{2}\right) \cos \frac{j\pi y}{b} \qquad \text{(A-3.51a)}$$

$$\overset{*}{B}_{kj} = \frac{4\phi_j \, \omega_k}{bm} \sum_{r=0}^{m-1} \int_0^b f(r,y) \cos \frac{k\pi}{m} \left(r + \frac{1}{2}\right) \cos \frac{j\pi y}{b} \, dy \qquad \text{(A-3.51b)}$$

Transformation from Infinite to Finite Series. – A problem frequently encountered in solving models which contain mixed discrete and continuous variables is that of finding a finite sinusoidal series to represent a discrete function that coincides at regularly spaced points with a continuous function that is known only in terms of the coefficients of an infinite sinusoidal series. For example, consider the case in which the continuous function is given as an infinite sine series.

$$f(x) \Big|_{x = \frac{a}{m} r} = \sum_{i=1}^{\infty} \overset{*}{A}_i \sin \frac{i\pi r}{m} = \sum_{k=1}^{m-1} A_k \sin \frac{k\pi r}{m} \qquad \text{(A-3.52)}$$

in which $r = 1,(1),m-1$ and it is required to find the finite set of A_k's in terms of the known infinite set of coefficients $\overset{*}{A}_i$. Routine use of the normal range (i.e., for $0 < k,k' < m$) expansion formulas, Eqs. A-3.35, leads to the result $A_k = \overset{*}{A}_k$ which is theoretically erroneous but is useful as an approximation when m is large and the infinite series converges rapidly. Inspection of the formula for A_k, Eq. A-3.35d, reveals that the coefficients are sine wise cyclic; i.e.,

$$A_k = A_{2Im+k} = -A_{2Im-k} = A_{k-2Im} \qquad \text{(A-3.53)}$$

$$\text{for} \quad I = -\infty, (1), +\infty$$

The applicable orthogonality conditions, Eq. A-3.35b, should be modified for an extended range of indices as follows:

(A-3.54a) $$\sin\frac{k\pi r}{m} = \sin\frac{\pi r}{m}(2Im + k) = -\sin\frac{\pi r}{m}(2Im - k)$$

(A-3.54b) $$\sum_{r=1}^{m-1}\sin\frac{i\pi r}{m}\sin\frac{k\pi r}{n} = \pm\frac{m}{2}\delta_j^{2Im \pm k}$$

i.e., the summation is zero except when $i = 2Im \pm k$. Use of this extended range orthogonality condition in the expansion formula (in addition to the identity $\overset{*}{A}_i = -\overset{*}{A}_{-i}$ from Eq. A-3.12b) yields the following alternate forms,

(A-3.55a) $$A_k = \overset{*}{A}_k - \overset{*}{A}_{2m-k} + \overset{*}{A}_{2m+k} - \overset{*}{A}_{4m-k} + \overset{*}{A}_{4m+k} + \ldots$$

(A-3.55b) $$A_k = \overset{*}{A}_k + \sum_{I=1}^{\infty}(\overset{*}{A}_{2Im+k} - \overset{*}{A}_{2Im-k})$$

(A-3.55c) $$A_k = \sum_{I=-\infty}^{+\infty}\overset{*}{A}_{2Im+k}$$

As an example, consider the transformation necessary to find the finite series coefficients for the node deflections of a simply supported uniformly loaded beam. The infinite series coefficients are

(A-3.56a) $$\overset{*}{A}_i = \frac{q_0}{B}\frac{4}{L}\left(\frac{L}{i\pi}\right)^5 \qquad i \text{ odd only}$$

(A-3.56b) $$A_k = q_0\frac{4L^4}{B\pi^5}\sum_{I=-\infty}^{+\infty}(2Im + k)^{-5} \simeq \frac{q_0}{B}\frac{4}{L}\left(\frac{L}{k\pi}\right)^5 \quad k \text{ odd only}$$

In this case the exact formal summation is available from a standard reference (e.g., through differentiation of Eq. 824, page 154, **Summation of Series**, by L.B.W. Jolley, Dover Publications, New York).

$$A_k = \frac{q_0}{12mB}\left(\frac{L}{m}\right)^4 \frac{6-\sigma_k}{\sigma_k^2} \cot\frac{k\pi}{2m} \qquad k \text{ odd only} \qquad (A\text{-}3.56c)$$

in which $\sigma_k = 1-\cos k\pi/m$. Due to the rapid convergence of the infinite series in this example, the truncated approximate formula in Eq. A-3.56b gives accurate numerical results. For $m = 6$ the approximate and exact answers for A_1 are, respectively, $A_1 \approx .01307106 \; q_0 L^4/B$ and $A_1 = .01307101 \; q_0 L^4/B$.

Analogous formulas can be written for coefficients of the finite cosine series.

$$f(x)\Big|_{x=\frac{a}{m}r} = \sum_{i=0}^{\infty} \phi_i \overset{*}{B}_i' \cos\frac{i\pi r}{m} = \sum_{k=0}^{m} B_k \cos\frac{k\pi r}{m} \qquad (A\text{-}3.57)$$

in which $r = 0,(1),m$.

From Eq. A-3.40e it is seen that the coefficients B_k are cosine wise cyclic for extended range indices; i.e.,

$$B_k = B_{2Im+k} = B_{2Im-k} = B_{k-2Im} \qquad (A\text{-}3.58a)$$

$$\sum_{r=0}^{m} \omega_r \cos\frac{i\pi r}{m} \cos\frac{k\pi r}{m} = \frac{m}{2\omega_k}\,\delta_j^{2Im \pm k} \qquad (A\text{-}3.58b)$$

in which $i = 0,(1),\infty$, $k = 0,(1),m$. The extended range orthogonality condition yields the following forms of the cosine transformation equation.

$$B_k = \omega_k\left[\overset{*}{B}_k' + \sum_{I=1}^{\infty} \left(\overset{*}{B}_{2Im+k}' + \overset{*}{B}_{2Im-k}'\right)\right] \qquad (A\text{-}3.59a)$$

$$B_k = \omega_k \sum_{I=-\infty}^{+\infty} \overset{*}{B}_{2Im+k}' \qquad (A\text{-}3.59b)$$

The required transformation formulae for coefficients of the finite displaced sine and cosine series are derived in a similar fashion. The results are:

$$f(x)\Big|_{x=\frac{a}{m}\left(r+\frac{1}{2}\right)} = \sum_{i=1}^{\infty} \overset{*}{A}_i \sin\frac{i\pi}{m}\left(r+\frac{1}{2}\right) = \sum_{k=1}^{m} A_k \sin\frac{k\pi}{m}\left(r+\frac{1}{2}\right)$$

(A-3.60a)

(A-3.60b) $$A_k = \phi_k \sum_{I=-\infty}^{+\infty} \overset{*}{A}_{2Im+k} (-1)^I \qquad k = 1,(1),m$$

$$f(x)\Big|_{x=\frac{a}{m}\left(r+\frac{1}{2}\right)} = \sum_{I=0}^{\infty} \phi_i \overset{*}{B}{}'_i \cos\frac{i\pi}{m}\left(r+\frac{1}{2}\right) = \sum_{k=0}^{m-1} B_k \cos\frac{k\pi}{m}\left(r+\frac{1}{2}\right)$$

(A-3.61a)

(A-3.61b) $$B_k = \omega_k \sum_{I=-\infty}^{+\infty} \overset{*}{B}{}'_{2Im+k} (-1)^I \qquad k = 0,(1), m-1$$

Extension of the above transformation formulae to multidimensional fields is accomplished routinely through repeated use of the analogous one-dimensional formulas; for example, consider the following two-dimensional case:

(A-3.62a) $$f(x,y)\Big|_{\substack{x=\frac{a}{m}r \\ y=\frac{b}{n}s}} = \sum_{i=0}^{\infty}\sum_{j=1}^{\infty} \phi_i \overset{*}{A}_{ij} \cos\frac{i\pi r}{m} \sin\frac{i\pi s}{n}$$

(A-3.62b) $$= \sum_{k=0}^{m}\sum_{\ell=1}^{n-1} A_{k\ell} \cos\frac{k\pi r}{m} \sin\frac{\ell\pi s}{n}$$

(A-3.62c) $$A_{k\ell} = \omega_k \sum_{I=-\infty}^{+\infty}\sum_{J=-\infty}^{\infty} \overset{*}{A}_{2Im+k,\,2Jn+\ell}$$

This concludes the Appendix on Fourier Series.

APPENDIX IV

MEMBRANE FORMULAS

Certain formulas from the classical plane stress elasticity solution for a rectangular membrane subjected to inplane loads and boundary displacements are required to account for composite action between the ribs and the plate in the analysis of an orthotropic deck.

Mathematical Model. – The equilibrium, force-deformation and governing displacement equations for a plane stress analysis of a rectangular continuum with inplane loads or body forces, f_x and f_y, are well known from elementary elasticity. For convenient reference and to define notation, the equations are shown below.

$$\begin{bmatrix} D_x & D_y & 0 \\ 0 & D_x & D_y \end{bmatrix} \begin{Bmatrix} n_x(x,y) \\ n_{xy}(x,y) \\ n_y(x,y) \end{Bmatrix} = \begin{Bmatrix} -f_x(x,y) \\ -f_y(x,y) \end{Bmatrix} \qquad \text{(A-4.1)}$$

in which D denotes differentiation with respect to the indicated subscript.

$$\begin{Bmatrix} n_x(x,y) \\ n_{xy}(x,y) \\ n_y(x,y) \end{Bmatrix} = K \begin{bmatrix} D_x & \mu D_y \\ \frac{1-\mu}{2} D_y & \frac{1-\mu}{2} D_x \\ \mu D_x & D_y \end{bmatrix} \begin{Bmatrix} u(x,y) \\ v(x,y) \end{Bmatrix} \qquad \text{(A-4.2)}$$

in which μ equals Poisson's ratio and $K = Et/(1 - \mu^2)$

$$\frac{1+\mu}{2} K \begin{bmatrix} \frac{2}{1+\mu} D_x^2 + \frac{1-\mu}{1+\mu} D_y^2 & D_x D_y \\ D_x D_y & \frac{1-\mu}{1+\mu} D_x^2 + \frac{2}{1+\mu} D_y^2 \end{bmatrix} \begin{Bmatrix} u(x,y) \\ v(x,y) \end{Bmatrix} = \begin{Bmatrix} -f_x(x,y) \\ -f_y(x,y) \end{Bmatrix}$$

(A-4.3)

$$u\left(x, {0 \atop b}\right) = n_y\left(x, {0 \atop b}\right) = 0 \qquad \text{(A-4a,b)}$$

Kernel Functions. – Particular solutions are required for the displacements due to a unit load at ξ, η in y direction (see Fig. A-1); that is, $f_x = 0$ and $f_y = \delta(y - \zeta)\delta(y - \eta)$ in which δ denotes the unit impulse load, a Dirac delta function in the continuous case. The displacement kernel functions are denoted $K^{uy}(x,y,\xi,\eta)$ and $K^{vy}(x,y,\xi,\eta)$. The series solutions are as follows:

$$K^{uy}(x,y,\xi,\eta) = \frac{4}{ab}\sum_{j=1}^{\infty}\sum_{i=1}^{\infty}\overset{*}{A}_{ij} \sin\alpha_i\xi \cos\bar{\alpha}_j\eta \cos\alpha_i x \sin\bar{\alpha}_j y$$
(A-4.5)

$$K^{vy}(x,y,\xi,\eta) = \frac{4}{ab}\sum_{j=0}^{\infty}\sum_{i=1}^{\infty}\overset{*}{\phi}_j \overset{*}{B}_{ij} \sin\alpha_i\xi \cos\bar{\alpha}_j\eta \sin\alpha_i x \cos\bar{\alpha}_j y$$
(4-4.6)

(A-4.7a,b,c) in which $\alpha_i = \frac{i\pi}{a}$, $\bar{\alpha}_j = \frac{j\pi}{b}$ and $\overset{*}{\phi}_j = 1 - \frac{1}{2}\delta_j^0$

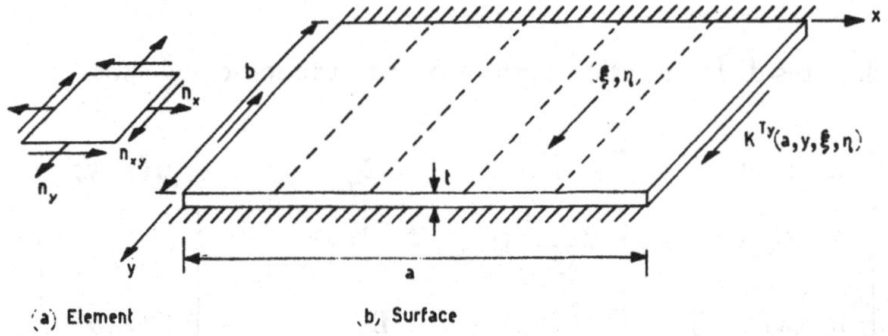

Fig. A-1. Membrane Action

(A-4.8)
$$\overset{*}{A}_{ij} = \left(-\frac{1}{K}\right)\frac{(1+\mu)\,\alpha_i\bar{\alpha}_j}{(1-\mu)(\alpha_i^2+\alpha_j^2)^2}$$

$$\overset{*}{B}_{ij} = \frac{1}{K}\,\frac{2\alpha_i^2+(1-\mu)\,\bar{\alpha}_j^2}{(1-\mu)(\alpha_i^2+\bar{\alpha}_j^2)^2}$$

The inplane membrane shear due to the above impulse loading is:

$$K^{Ty}(x,y,\xi,\eta) = \frac{4}{ab}\sum_{j=0}^{\infty}\sum_{i=1}^{\infty}\overset{*}{\phi}_j \overset{*}{\bar{B}}_{ij} \sin\alpha_i\xi \cos\bar{\alpha}_j\eta \cos\alpha_i x \cos\bar{\alpha}_j y$$
(A-4.10)

$$\overset{*}{\bar{B}}_{ij} = \frac{\alpha_i(\alpha_i^2 - \mu\bar{\alpha}_j^2)}{(\alpha_i^2+\bar{\alpha}_j^2)^2}$$
(A-4.11)

Rib Line Solutions: For orthotropic deck analysis, many of the membrane quantities are required at the rib lines. This necessitates transformation of the infinite series on x to finite series on r. The techniques for such transformations are shown in reference 3. The results of these transformations are as follows:

$$K^{uy}(r,y,\alpha,\eta) = \frac{4}{mb}\sum_{j=1}^{\infty}\sum_{k=1}^{m-1} A_{kj} \sin\frac{k\pi\alpha}{m} \cos\bar{\alpha}_j\eta \cos\frac{k\pi r}{m} \sin\bar{\alpha}_j y \tag{A-4.12}$$

$$K^{vy}(r,y,\alpha,\eta) = \frac{4}{mb}\sum_{j=0}^{\infty}\sum_{k=1}^{m-1}{}^{*} \phi_j B_{kj} \sin\frac{k\pi\alpha}{m} \cos\bar{\alpha}_j\eta \sin\frac{k\pi r}{m} \cos\bar{\alpha}_j y \tag{A-4.13}$$

$$A_{kj} = \frac{-a(1+\mu)}{4mK(1-\mu)\bar{D}_{kj}^2} \sin\frac{k\pi}{m} \sinh\lambda_j \tag{A-4.14}$$

$$B_{kj} = \frac{a}{4mK(1-\mu)\bar{D}_{kj}}\left[\frac{3-\mu}{\lambda_j}\sinh\lambda_j + \frac{1+\mu}{\bar{D}_{kj}}\left(1-\cosh\lambda_j \cos\frac{k\pi}{m}\right)\right] \tag{A-4.15}$$

$$\lambda_j = \frac{a}{m}\bar{\alpha}_j \qquad \bar{D}_{kj} = \cosh\lambda_j - \cos\frac{k\pi}{m} \qquad B_{kj} \simeq \frac{m}{a}\overset{*}{B}_{kj} \tag{A-4.16}$$

$$K^{Ty}(r,y,\alpha,\eta) = \frac{4}{mb}\sum_{j=0}^{\infty}\sum_{k=1}^{m-1}{}^{*} \phi_j \bar{B}_{kj} \sin\frac{k\pi\alpha}{m} \cos\bar{\alpha}_j\eta \cos\frac{k\pi r}{m} \cos\bar{\alpha}_j y \tag{A-4.17}$$

$$\bar{B}_{kj} = \frac{\sin\frac{k\pi}{m}}{4\bar{D}_{kj}}\left[2 - \frac{(1+\mu)\lambda_j \sinh\lambda_j}{\bar{D}_{kj}}\right] \tag{A-4.18}$$

Rib Formulas. – The beam model for transverse and longitudinal loads applied at the top edge of a deck rib – the portion below the continuous deck surface (see Fig. A-2) – is as follows:

$$AE\begin{bmatrix} (\rho^2 + e^2)D_y^4 & -eD_y^3 \\ eD_y^3 & -D_y^2 \end{bmatrix}\begin{Bmatrix} w(y) \\ v(y) \end{Bmatrix} = \begin{Bmatrix} N(y) \\ -F(y) \end{Bmatrix} \tag{A-4.19}$$

$$w\binom{0}{b} = D_y^2 w\binom{0}{b} = D_y v\binom{0}{b} = 0 \tag{A-4.20}$$

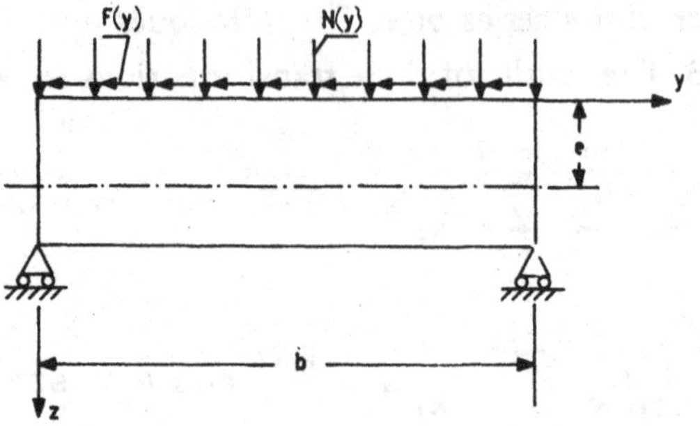

Fig. A-2. Rib Forces

in which A, E and ρ are the area, modulus of elasticity and radius of gyration respectively, e is the eccentricity of the applied shear load from the centroid, N(y) and F(y) are the applied transverse and shear loads and w(y) and v(y) are the top edge displacements in the y and z directions respectively.

The required solution is a set of kernel functions for a diagonal set of impulse loads; that is, w(y) and v(y) for $N(y) = \delta(y-\eta)$ and $F(y) = 0$, $K^{wz}(y,\eta)$, and $K^{vz}(y,\eta)$, and w(y) and v(y) for $N(y) = 0$ and $F(y) = \delta(y-\eta)$, $K^{wy}(y,\eta)$ and $K^{vy}(y,\eta)$. These definitions and the kernel function formulas are shown in the following equations.

$$\begin{Bmatrix} N(y) \\ F(y) \end{Bmatrix} = \begin{Bmatrix} \delta(y-\eta) & 0 \\ 0 & \delta(y-\eta) \end{Bmatrix} ; \begin{Bmatrix} w(y) \\ v(y) \end{Bmatrix} = \begin{bmatrix} K^{wz}(y,\eta) & K^{wy}(y,\eta) \\ K^{vz}(y,\eta) & K^{vy}(y,\eta) \end{bmatrix}$$

(A-4.21)

$$\begin{bmatrix} K^{wz}(y,\eta) & K^{wy}(y,\eta) \\ K^{vz}(y,\eta) & K^{vy}(y,\eta) \end{bmatrix} = \frac{2}{b}\sum_{j=1}^{\infty} \begin{bmatrix} \overset{*}{A}_j \sin\bar{\alpha}_j\eta \sin\bar{\alpha}_j y & -\overset{*}{D}_j \cos\bar{\alpha}_j\eta \sin\bar{\alpha}_j y \\ \overset{*}{D}_j \sin\bar{\alpha}_j\eta \cos\bar{\alpha}_j y & -\overset{*}{B}_j \cos\bar{\alpha}_j\eta \cos\bar{\alpha}_j y \end{bmatrix}$$

(A-4.22)

(A-4.23) $$\overset{*}{A}_j = \frac{1}{B\bar{\alpha}_j^4} ; \quad \overset{*}{D}_j = \frac{e}{B\bar{\alpha}_j^3} ; \quad \overset{*}{B}_j = \frac{\rho^2 + e^2}{B\bar{\alpha}_j^2}$$

in which $B = EA\rho^2$, the flexural rigidity of the rib. (Note that the term including $\overset{*}{B}_0$ is omitted due to the fact that F(y) will be self equilibrating.)

This completes the required set of solutions for the rib beams.

APPENDIX V

FLEXURAL PLATE FORMULAS

Certain formulas from classical plate theory, and their extension to express rib line descriptors, are required to include the out-of-plane plate stiffness effects in the analysis of such mixed discrete-continuous systems as waffle or reinforced building slabs and bridge decks. Such formulas are available, see Ref. 100, but are duplicated below for convenience.

(A-5.1) $$[\,D_x^2,\ 2D_xD_y,\ D_y^2\,]\begin{Bmatrix} m_x(x,y) \\ m_{xy}(x,y) \\ m_y(x,y) \end{Bmatrix} = -\,q(x,y)$$

(A-5.2) $$\begin{Bmatrix} m_x(x,y) \\ m_{xy}(x,y) \\ m_y(x,y) \end{Bmatrix} = -\,D \begin{Bmatrix} D_x^2 + \mu D_y^2 \\ (1-\mu)D_xD_y \\ \mu D_x^2 + D_y^2 \end{Bmatrix} w(x,y)$$

(A-5.3) $$(D_x^2 + D_y^2)^2\, w(x,y) = \frac{1}{D}\, q(x,y)$$

(A-5.4) $$w\left(x, {0 \atop b}\right) = m_y\left(x, {0 \atop b}\right) = 0$$

(A-5.5) $$D = \frac{Et^3}{12(1-\mu^2)}$$

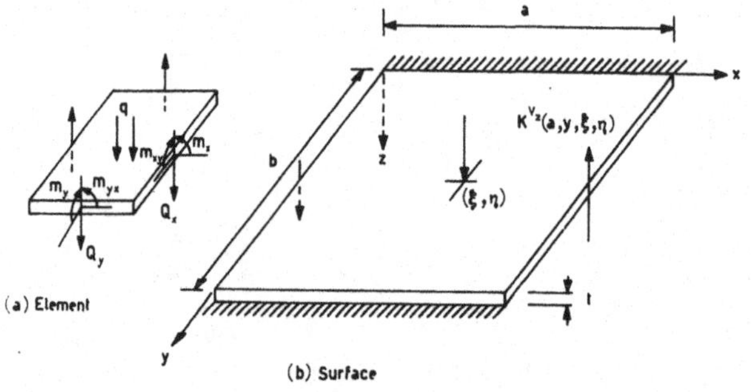

Fig. A-3. Flexural Action

Kernel Functions. – The deflection of a simply supported plate at coordinates x,y due to an out-of-plane unit concentrated load (impulse load) at coordinates ξ, η – termed a kernel function – is :

$$K^{wz}(x,y,\xi,\eta) = \frac{4}{ab}\sum_{j=1}^{\infty}\sum_{i=1}^{\infty}\overset{*}{C}_{ij}\ \sin\alpha_i\xi\ \sin\bar{\alpha}_j\eta\ \sin\alpha_i x\ \sin\bar{\alpha}_j y \tag{A-5.6}$$

$$\overset{*}{C}_{ij} = \frac{1}{D(\alpha_i^2+\bar{\alpha}_j^2)^2}\ ;\ \alpha_i = \frac{i\pi}{a}\ ;\ \bar{\alpha}_j = \frac{j\pi}{b} \tag{A-5.7}$$

For a distributed plate loading q(x, y) the displacement field is as follows :

$$w^a(x,y) = \sum_{j=1}^{\infty}\sum_{i=1}^{\infty}\overset{*}{C}_{ij}\ \overset{*}{q}_{ij}\ \sin\alpha_i x\ \sin\bar{\alpha}_j y \tag{A-5.8}$$

$$q(x,y) = \sum_{j=1}^{\infty}\sum_{i=1}^{\infty}\overset{*}{q}_{ij}\ \sin\alpha_i x\ \sin\bar{\alpha}_j y \tag{A-5.9}$$

$$\overset{*}{q}_{ij} = \frac{4}{ab}\int_0^a\int_0^b q(x,y)\ \sin\alpha_i x\ \sin\bar{\alpha}_j y\ dx\ dy \tag{A-5.10}$$

The shear resultant, physically identified only at the boundaries x = 0,a, is given by

$$K^{Vz}(x,y,\xi,\eta) = -D\,D_x[D_x^2 + (2-\mu)D_y^2]\,K^{wz}(x,y,\xi,\eta) \tag{A-5.11}$$

$$K^{Vz}(x,y,\xi,\eta) = \frac{4}{ab}\sum_{j=1}^{\infty}\sum_{i=1}^{\infty}\overset{*}{\bar{C}}_{ij}\ \sin\alpha_i\xi\sin\bar{\alpha}_j\eta\ \cos\alpha_i x\ \sin\bar{\alpha}_j y \tag{A-5.12}$$

$$\overset{*}{\bar{C}}_{ij} = \frac{\alpha_i[\alpha_i^2 + (2-\mu)\bar{\alpha}_j^2]}{(\alpha_i^2+\bar{\alpha}_j^2)^2} \tag{A-5.13}$$

Imposed Boundary Displacements. – Homogeneous flexural plate solutions (q(x, y) = 0) are required for decks with nonzero out-of-plane side boundary displacements.

$$w^h\left({0 \atop a}, y\right) = \sum_{j=1}^{\infty}(W_j^{\$} \pm W_j^{a/s})\ \sin\bar{\alpha}_j y \tag{A-5.14}$$

$$w^h(x,y) = \frac{4}{a}\sum_{j=1}^{\infty}\sum_{i=1}^{\infty}\bar{W}_{ij}\ \overset{*}{\bar{C}}_{ij}\ \sin\alpha_i x\ \sin\bar{\alpha}_j y \tag{A-5.15}$$

$$\bar{W}_{ij} = \begin{bmatrix} W_j^{\$} & \text{for } i \text{ odd} \\ W_j^{a/s} & \text{for } i \text{ even}\end{bmatrix} \tag{A-5.16}$$

As with the membrane solution, the solution form in Eq. A-5.15 converges slowly in some cases and should be replaced by the following mixed form

$$w^{h}(x,y) = \sum_{j=1}^{\infty}\left[W_j^{\$} + \left(1-\frac{2x}{a}\right)W_j^{a/s} + \frac{4}{a}\sum_{i=1}^{\infty}\bar{W}_{ij}\left(\bar{\bar{C}}_{ij}^{*} - \frac{1}{\alpha_i}\right)\sin\alpha_i x\right]\sin\bar{\alpha}_j y$$

(A-5.17)

Note that the algebraic expression in Eq. A-17 is valid everywhere except for the moment boundary condition ; $m_x\left({}^{0}_{a}, y\right) = 0$. This form is computationally superior and simpler than the exact boundary function, used in deriving Eq. A-15, and shown below ;

$$\left[1-\frac{\mu}{2}\bar{\alpha}_j^2 x(a-x)\right]W_j^{\$} + \left(1-\frac{2x}{a}\right)\left[1-\frac{\mu}{6}\bar{\alpha}_j^2 x(a-x)\right]W_j^{a/s}$$

The shear resultants, due to imposed boundary displacements, to be evaluated along the boundaries, are :

(A-5.18) $$V_x^h(x,y) = \frac{4}{a}\sum_{j=0}^{\infty}\sum_{i=0}^{\infty}\phi_i^{*}\bar{S}_{ij}\cos\alpha_i x \sin\bar{\alpha}_j y$$

(A-5.19) $$\bar{S}_{ij}^{*} = -(1-\mu)D\frac{\bar{\alpha}_j^4\left[(1+\mu)\alpha_i^2 + 2\bar{\alpha}_j^2\right]}{(\alpha_i^2+\bar{\alpha}_j^2)^2}$$

(A-5.20) $$V_x^h\left({}^{0}_{a}, y\right) = \sum_{j=1}^{\infty}\left(W_j^{a/s} S_j^{h\,a/s} \pm W_j^{\$} S_j^{h\$}\right)\sin\bar{\alpha}_j y$$

(A-5.21) $$\begin{bmatrix} S_j^{h\$} \\ \\ S_j^{h\,a/s}\end{bmatrix} = \frac{-(1-\mu)}{2}D\bar{\alpha}_j^3\left[\frac{(3+\mu)\sinh a\bar{\alpha}_j \mp (1-\mu)a\bar{\alpha}_j}{\cosh a\bar{\alpha}_j \pm 1}\right]$$

(A-5.22) $$\begin{bmatrix} S_j^{h\$} \\ \\ S_j^{h\,a/s}\end{bmatrix} \cong \frac{4}{a}\begin{bmatrix}\bar{S}_{ij}^{*} \\ \\ \frac{1}{2}\bar{S}_{0j}^{*}\end{bmatrix}$$

Rib Line Solutions. – As was the case for the membrane formulas, several of the continuous flexural plate solutions are required in terms of discrete-continuous coordinates to evaluate the dependent variable along an arbitrary rib line. The required expressions are as follows :

$$K^{wz}(r,y,\alpha,\eta) = \frac{4}{mb}\sum_{j=1}^{\infty}\sum_{k=1}^{m-1} C_{kj} \sin\frac{k\pi\alpha}{m} \sin\bar{\alpha}_j\eta \sin\frac{k\pi r}{m} \sin\bar{\alpha}_j y \tag{A-5.23}$$

$$C_{kj} = \frac{a}{4mD\,\bar{\alpha}_j^2\,\bar{D}_{kj}}\left[\frac{\sinh\lambda_j}{\lambda_j} - \frac{1-\cosh\lambda_j\cos\frac{k\pi}{m}}{\bar{D}_{kj}}\right] \tag{A-5.24}$$

$$C_{kj} \simeq \frac{m}{a}\overset{*}{C}_{kj} \tag{A-5.25}$$

in which λ_j and $\bar{D}_{kj}$ are given by Eq. A-4.16

$$K^{Vz}(r,y,\alpha,\eta) = \frac{4}{mb}\sum_{j=1}^{\infty}\sum_{k=1}^{m-1} \bar{C}_{kj} \sin\frac{k\pi\alpha}{m} \sin\bar{\alpha}_j\eta \cos\frac{k\pi r}{m} \sin\bar{\alpha}_j y \tag{A-5.26}$$

$$\bar{C}_{kj} = \frac{\sin\frac{k\pi}{m}}{4\bar{D}_{kj}}\left[2 + \frac{(1-\mu)\lambda_j\sinh\lambda_j}{\bar{D}_{kj}}\right] \tag{A-5.27}$$

(use of the approximation $\bar{C}_{kj} \simeq m/a\ \overset{*}{\bar{C}}_{kj}$ is not recommended due to insufficient accuracy).

$$w^a(r,y) = \sum_{j=1}^{\infty}\sum_{k=1}^{m-1} W^a_{kj} \sin\frac{k\pi r}{m} \sin\bar{\alpha}_j y \tag{A-5.28}$$

$$W^a_{kj} = \sum_{I=-\infty}^{+\infty} \overset{*}{C}_{2Im+k}\,\overset{*}{q}_{2Im+k,j} \simeq \overset{*}{C}_{kj}\,\overset{*}{q}_{kj} \tag{A-5.29}$$

$$w^h(r,y) = \frac{4}{m}\sum_{j=1}^{\infty}\sum_{k=1}^{m-1} \bar{W}_{kj}\,\bar{C}_{kj} \sin\frac{k\pi r}{m} \sin\bar{\alpha}_j y \tag{A-5.30}$$

The boundary shear resultants due to applied loads with simple side supports are :

$$(\text{A-5.31}) \qquad V_x^a\left({0 \atop a}, y\right) = \sum_{j=1}^{\infty} (S_j^{a\,a/s} \pm S_j^{a\,\$}) \sin \bar{\alpha}_j\, y$$

in which

$$(\text{A-5.32}) \qquad \begin{bmatrix} S_j^{a\,\$} \\ \\ S_j^{a\,a/s} \end{bmatrix} = \sum_i \overset{*}{\bar{C}}_{ij} \, \overset{*}{q}_{ij} \qquad \text{for} \quad \begin{bmatrix} i = 1,(2),\infty \\ \\ i = 2,(2),\infty \end{bmatrix}$$

CONTENTS

Printed in the United States
By Bookmasters